Wittgenstein on Practice

Kevin M. Cahill
Editor

Wittgenstein on Practice

Back to the Rough Ground

palgrave
macmillan

Editor
Kevin M. Cahill
Department of Philosophy
University of Bergen
Bergen, Norway

ISBN 978-3-031-68655-9   ISBN 978-3-031-68656-6 (eBook)
https://doi.org/10.1007/978-3-031-68656-6

This Palgrave Macmillan imprint is published by the registered company Springer Nature Switzerland AG.
The registered company address is: Gewerbestrasse 11, 6330 Cham, Switzerland

Acknowledgements

Work on this volume was carried out with generous support from the Norwegian Research Council as part of the project "Mathematics with a Human Face: Set Theory within a Naturalized Wittgensteinian Framework". I am also grateful to my colleague and project leader Professor Sorin Bangu for his support and encouragement. Finally, I'd like to thank the contributors to this collection.

Contents

Notes on Contributors

Reshef Agam-Segal is Professor of Philosophy at the Virginia Military Institute, USA. His main areas of research are Wittgenstein's philosophy and ethics, and he has a special interest in aspect-perception and its connection to several philosophical domains, including the philosophy of mind, the philosophy of mathematics, the philosophy of perception, and ethics. He is the co-editor of *Wittgenstein's Moral Thought* (Routledge, 2018).

Valérie Aucouturier is Professor of Philosophy at UCLouvain Saint-Louis Brussels, Belgium, and a member of the *Centre Prospéro. Langage, Image, Connaissance*. Her research mainly focuses on the philosophy of language, mind, and action (Anscombe, Austin, Wittgenstein) and the philosophy of psychology and psychoanalysis. She has published extensively on the work of Elizabeth Anscombe and philosophy of mind, action, and language. Her main publications include *Elizabeth Anscombe, l'esprit en pratique* (CNRS editions, 2012) and *L'intention en action* (Vrin, 2018).

Kim-Erik Berts is University Lecturer in Mathematics Education at Åbo Akademi University, Finland. He holds a PhD in Philosophy. His research interests center around the concepts of mathematical proof and argumentation and how certainty is established within mathematics, especially for the learner. He has published on the nature of mathematical proof and on Wittgenstein's philosophy of mathematics.

Kevin M. Cahill is Professor of Philosophy at the University of Bergen, Norway. His primary areas of research are in Wittgenstein studies and in the philosophy of the social sciences. In addition to several journal articles and book chapters, his main publications include *The Fate of Wonder: Wittgenstein's Critique of Metaphysics and Modernity* (Columbia, 2011), *Wittgenstein and Naturalism* (co-edited with Thomas Raleigh) (Routledge, 2018), and *Towards a Philosophical Anthropology of Culture* (Routledge, 2021).

Juliet Floyd is Borden Parker Bowne Professor of Philosophy at Boston University and director of the Boston University Center for the Humanities. She has written many articles on the history and development of twentieth-century analytic and American philosophy, philosophy of logic and mathematics, ordinary language, symbolic forms, aesthetics, and philosophies of computational and emerging media. Her most recent books are *Wittgenstein's Philosophy of Mathematics* (Cambridge, 2021) and *Nudging Choices Through Media: Ethical and Philosophical Implications for Humanity*, co-edited with J.E. Katz and K. Schiepers (Springer/Palgrave, 2023).

Martin Gustafsson is Professor of Philosophy at Åbo Akademi University, Finland. His main research areas are the philosophy of language and the philosophy of action. He is particularly interested in how they hang together—the ways in which language and action intermingle and are embedded in the wider circumstances of human life. Gustafson's research areas also include the history of analytic philosophy, philosophy of logic, philosophical methodology, and the philosophy of the human sciences.

Lars Hertzberg is Professor Emeritus of Philosophy at Åbo Akademi University, Finland. He has written essays on ethics, the philosophy of language, philosophical psychology, and Wittgenstein, some of them collected in *The Limits of Experience* (Acta Philosophical Fennica, 1994) and in *Wittgenstein and the Life We Live with Language* (Anthem Press, 2022). He has also written *Wittgenstein on Criteria and Practices* (Cambridge Element, 2022).

Oskari Kuusela is an associate professor at the University of East Anglia. His monographs include *The Struggle Against Dogmatism* (HUP 2008) and *Wittgenstein on Logic as the Method of Philosophy* (OUP 2019). He is the co-editor of five collections on Wittgenstein, including the *Oxford Handbook of Wittgenstein* (OUP 2011) and most recently *Engaging Kripke with Wittgenstein* (Routledge 2023).

Cheryl Misak is University Professor and Professor of Philosophy at the University of Toronto. She works on American pragmatism, the history of analytic philosophy, ethics and political philosophy, and the philosophy of medicine. Her books include *Frank Ramsey: A Sheer Excess of Powers*, *Cambridge Pragmatism*, *The American Pragmatists*, *Truth and the End of Inquiry*, and *Truth, Politics, Morality.*

Mark Risjord is Professor of Philosophy at Emory University where he directs the Institute for the Liberal Arts. Risjord's research focuses on the philosophy of science, particularly the social and biomedical sciences, including nursing. He has published many essays, and his books include *Woodcutters and Witchcraft: Rationality and Interpretive Change in the Social Sciences* (SUNY Press 2000), *Nursing Knowledge: Science, Practice, and Philosophy* (Wiley-Blackwell, 2010), and *A Contemporary Introduction to the Philosophy of Social Science* (Routledge, 2023, second edition).

Joseph Rouse is the Hedding Professor of Moral Science in the Department of Philosophy and the College of Science and Technology Studies at Wesleyan University. His primary research areas are the philosophy of scientific practice, the social theory of practices, the philosophy of biology, and naturalism in philosophy. He is the author of *Social Practices as Biological Niche Construction* (Chicago 2023), *Articulating the World* (Chicago 2015), *How Scientific Practices Matter* (Chicago 2002), *Engaging Science* (Cornell 1996), and *Knowledge and Power* (Cornell 1987), and editor of John Haugeland's posthumous *Dasein Disclosed.*

Nancy Yousef is Professor of English at Rutgers University, New Brunswick. Her work explores intersections between philosophical writing and literary form, aesthetics, and ethics. She is the author of *Isolated Cases: The Anxieties of Autonomy from Enlightenment to Romanticism* (Cornell, 2004), *Romantic Intimacy* (Stanford, 2013), and *The Aesthetic Commonplace: Wordsworth, Eliot, Wittgenstein and the Language of Everyday* (Oxford, 2022).

Abbreviations[1] of Wittgenstein's Works

BBB	The Blue and the Brown Books
BEE	Wittgenstein's *Nachlass*: The Bergen Electronic Edition
CV	Culture and Value
LC	Lectures and Conversations on Aesthetics, Psychology and Religious Belief
LFM	Lectures on the Foundations of Mathematics
OC	On Certainty
PG	Philosophical Grammar
PI	Philosophical Investigations
PO	Philosophical Occasions
PPO	Ludwig Wittgenstein: Public and Private Occasions
PR	Philosophical Remarks
RFGB	Remarks on Frazer's *Golden Bough* (in PO)
RFM	Remarks on the Foundations of Mathematics
TLP	Tractatus Logico-Philosophicus
WML	Wittgenstein: Lectures, Cambridge 1930-1933, From the Notes of G.E. Moore
Z	Zettel

[1] See the respective chapters for bibliographic information on specific editions.

Introduction

Kevin M. Cahill

The complexity of our current philosophical landscape makes it difficult to imagine that any unifying thinker, movement, or philosophical work will emerge any time soon to unite the discipline, either in perspective or concern. Things have simply become too messy for any sort of well-ordered philosophical cosmos. Still, there are commonalities of interest that for the time being provide the discipline of philosophy with something like a manageable map. One such concentration is shared by philosophers who believe that philosophical reflection on topics ranging from mathematics, biology, ethics, and language is at its best when informed by considering these subjects as involving human, or at least creaturely, activities. While such a perspective does not require that one be a Wittgensteinian, there is little doubt that Wittgenstein's work has provided a major impetus for this direction in contemporary philosophy. Few contemporary pragmatists, practice theorists, phenomenologists, or philosophical naturalists would be willing to reject this aspect of his

K. M. Cahill (✉)
Department of Philosophy, University of Bergen, Bergen, Norway
e-mail: kevin.cahill@uib.no

K. M. Cahill (ed.), *Wittgenstein on Practice*,
https://doi.org/10.1007/978-3-031-68656-6_1

thought, and many, without regarding themselves as out-and-out Wittgensteinians, are glad to draw inspiration from his work. This includes such thinkers as Robert Brandom, Toril Moi, John McDowell, Hew Price, Martha Nussbaum, Charles Taylor, and numerous others. As of yet, however, there is no existing volume of essays dedicated to the theme of Wittgenstein and practice. Next is a brief sketch of my own view on the meaning of practice in Wittgenstein's later work. For simplicity's sake, I focus here exclusively on the *Philosophical Investigations* (*PI*). This is followed by a precis of each of the chapters gathered here.

In the remark immediately preceding his introduction of the concept of "family resemblance" in *PI*, Wittgenstein imagines himself in dialogue with a philosopher insisting that a substantive like "game" must have an essential common property, for otherwise we could not use the term correctly when applied to all of its various instantiations. Wittgenstein follows a statement of the essentialist thesis with the now familiar admonishment "Don't think, but look!

> Look for example at board-games, with their multifarious relationships. Now pass to card-games; here you find many correspondences with the first group, but many common features drop out, and others appear. When we pass next to ball-games, much that is common is retained, but much is lost.—Are they all 'amusing'? Compare chess with noughts and crosses. Or is there always winning and losing, or competition between players? Think of patience. In ball games there is winning and losing; but when a child throws his ball at the wall and catches it again, this feature has disappeared. Look at the parts played by skill and luck; and at the difference between skill in chess and skill in tennis. Think now of games like ring-a-ring-a-roses; here is the element of amusement, but how many other characteristic features have disappeared! And we can go through the many, many other groups of games in the same way; can see how similarities crop up and disappear.
>
> And the result of this examination is: we see a complicated network of similarities overlapping and criss-crossing: sometimes overall similarities, sometimes similarities of detail. (PI §66)

Forty remarks on, Wittgenstein is still grappling with his essentialist interlocutor, but now the focus has shifted somewhat to the purported sublime nature of logic. We read at *PI* §107,

> The more narrowly we examine actual language, the sharper becomes the conflict between it and our requirement. (For the crystalline purity of logic was, of course, not a *result of investigation*: it was a requirement.) The conflict becomes intolerable; the requirement is now in danger of becoming empty.—We have got on to slippery ice where there is no friction and so in a certain sense the conditions are ideal, but also, just because of that, we are unable to walk. We want to walk: so we need *friction*. Back to the rough ground!

The rough ground (aka the mouse and the rags of §52), the philosopher assumes to be so philosophically uninteresting as not even to warrant a look, is of course our life with language: in other words the uses (*Gebrauch*), customs (*Gepflogenheiten*), or techniques (*Techniken*) of signs. At the risk of running over some important distinctions, I am going to treat "practice" (*Praxis*) here as an abbreviation standing both for itself and for these related terms. These concepts are surely important; perhaps, then, what the later Wittgenstein brought to the table that was original resides in his groundbreaking discovery of the relevance of human practice for philosophy. Yet this is potentially quite misleading.

Far from being a novel idea, the idea that practice, or notions very much like it, have philosophical significance goes as far back as Aristotle and can be found in the work of such venerable figures as Aquinas, Kant, Hegel, Marx, Nietzsche, Peirce, James, and Dewey, existential phenomenologists like Heidegger and Merleau-Ponty, even the early Wittgenstein himself (See TLP 3.327). Any claim to originality and profundity for the later Wittgenstein's thought, therefore, cannot rest on his merely having drawn our attention to the facts of the upbringing and training of human beings.

So where does the later Wittgenstein's contribution to our thinking about practice lie? I will emphasize two elements here.

First, there is his emphasis on the *logical* significance of our grammatical practices, broadly understood. Wittgenstein's investigations of what we (are taught to) say and when we are to say it are more directly oriented towards meaning than towards truth. I should immediately add that in stressing Wittgenstein's preoccupation with meaning and conceptual matters in this way, I shouldn't be taken to be suggesting that he had no interest in truth. But I am saying that whatever concern with truth he may have had, as with any of his other philosophical concerns, was always mediated by his concern with grammatical practices. (This is no place to deal with the question of where I think this places Wittgenstein in realist/anti-realist debates.) Moreover, these practices, in turn, don't so much as determine meaning as they provide a logical shape for the space in which meaning is found.

Now using the expressions "shape" and "space" in this way may make it seem as though practices, understood broadly, provide a kind of bounded framework for meaning, perhaps even a foundation (A kind of view that Oskari Kuusela argues is a serious misreading.) Wittgenstein believes that in our puzzlements about meaning, we either won't consider our practices, or, if we do, then we are tempted to adopt a ham-fisted equation such as "meaning is use". A better maxim would be, "Look at the use and you (might well) see the meaning." My talk of the logical significance of our grammatical practices residing in their providing a logical shape for the space in which meaning is found is pretty indeterminate. There is no getting around that.

The second element I want to mention here, and for some readers this will only make matters even more intolerably vague, is how Wittgenstein connects our practice-infused activities with our creaturely nature. These connections are made in several numerous remarks concerning our "primitive, …natural expressions" (§244, §256, §647), our "natural history" (§25, §415, "very general facts of nature, p. 231), and of course in somewhat notorious passages where he speaks of "form(s) of life" (§19, §23, §241, p. 174, p. 226). As with practice, taken as broadly defined above, none of these passages are invitations to understand Wittgenstein as offering a biological explanation of meaning and understanding or as providing meaning and understanding with a biological foundation. The best we can say is that he is gesturing at, if I can put it this way, very unschematized "conditions of the possibility" of meaning, with no pretense to a prioricity.

Wittgenstein on Practice: Back to the Rough Ground brings together twelve previously unpublished chapters. While Wittgenstein's thought frames the collection as a whole, each contribution aims first and foremost at rigorous philosophical argument directed at contemporary issues. In this sense, each contribution "drafts" Wittgenstein on practice either by following in his wake, by sketching or critiquing some aspect of his thought, or both. The shared element is careful attention to human (or animal) activity as essential to thinking about philosophical problems.

An End to Skimming (Yousef)

To a certain extent, Nancy Yousef's "On the Use and Misuse of Wittgenstein for Literature or, The (Hard) Practice of Reading" stands apart, not merely because Yousef is not a professional philosopher, but mainly because it is the most meta-philosophical and interdisciplinary contribution to the volume. This makes it all the more appropriate that Yousef's chapter be included first here. Wittgenstein's work is enjoying a modest surge of interest among literary scholars advocating a turn away from the "ideological critique" and "hermeneutics of suspicion" that have been the dominant paradigms for critical practice in the literary humanities for the past half-century, especially in the Anglo-American academy. While there have always been dissident counter-strains, several recent high-profile theoretical polemics—most notably Moi (2017) and Zumhagen-Yekplé (2020)—have championed Wittgenstein (and "ordinary language philosophy" more broadly) as a resource for an approach to literary interpretation that would be both ethically transformative and "post-critical." A touchstone for these new Wittgensteinian critics is the notion of attunement to the near-at-hand yet overlooked, as in *Philosophical Investigations* §129: "The aspects of things that are most important for us are hidden because of their simplicity and familiarity. (One is unable to notice something—because it is always before one's eyes.")

Yousef aims to offer an alternative account of how Wittgenstein may provide a model and resource for literary studies. Her essay explores Wittgenstein's methods of working and writing as entailing also a reading practice of heightened attention to the affective implications of small and subtle shifts in wording and context, the devoted labor of repetition and revision, hence the extraordinary commitment to revision and rewriting evidenced in the *Nachlass*. The greater part of the essay offers a case study of Wittgenstein's multiyear working through and working over the famous remark, "The human body is the best picture of the human soul" (*PPF* iv §25). The paper speaks to the challenges of genuine interdisciplinary engagement with philosophy from the side of literary studies, but also shows that the attention to the aesthetic resources of language (and especially to figurative language) central to literary criticism may have something to contribute to efforts, on the side of philosophy, to revive a substantive practice of engagement with the texts comprising its own history (what James Conant calls "philosophically illuminating history of philosophy").

Learning to Go On (Hertzberg, Aucouterier, Kuusela, and Gustafsson)

Since the publication of Saul Kripke's *Wittgenstein on Rules and Private Language*, a rich literature has evolved that addresses both Kripke's interpretation of Wittgenstein's remarks on rule-following in *Philosophical Investigations* and of course the text of the remarks themselves. Even philosophers who, like Kripke himself, would shun the label "Wittgensteinian" acknowledge that the remarks raise utterly central issues concerning meaning and understanding. There is also widespread agreement that such related notions of practice and custom are key to understanding Wittgenstein's overall point. After all, at *Philosophical Investigations* §202, which can be considered the punchline to this series of remarks, Wittgenstein tells us that "'following a rule' is a practice."

In his essay, Lars Hertzberg examines how this remark should be understood. Which specific kinds of practice does Wittgenstein have in mind? The word "practice" is not used frequently in *PI*, although there are numerous remarks that can be held to bear on the concept. Hardly any concrete examples are given, however, apart from the playing of games which is, in any case, not a typical form of practice. In *Zettel* §320 Wittgenstein distinguishes between rules that are, in a sense, arbitrary (rules of grammar, chess) and rules that are not (cookery rules) and Hertzberg argues that this dichotomy is problematic. He goes on to look at some concrete practices, such as greeting, and discuss the different senses in which they can be said to embody rules.

In "Wittgenstein and Kripke on Rule-Following: The Problem of Empty Philosophical Explanations," Oskari Kuusela poses questions about the kind of explanation many interpreters assume to be offered when Wittgenstein says that rule-following is a practice or, as in Kripke's influential exposition, that rule-following is based on communal agreement. Kuusela finds such readings lacking and shows why on closer inspection they turn out to be empty. They constitute merely apparent explanations that can't do the explanatory job they are intended to do. Consequently, the role of practice as part of a grammatical clarification of what rule-following is must be understood differently. Kuusela outlines an account that he believes to be a better fit with Wittgenstein's philosophical approach and methodological commitments.

Valérie Aucouturier's "Wittgenstein, Meaning and Action" seeks to clarify the relation between meaning and action in Wittgenstein's later thought. According to Aucouturier, two problems are intertwined here: First, if following a rule can only be described from *within* a practice, how can we identify a rule independently of individual acts of rule-following? Secondly, if genuine rule-following is unintelligible apart from our actions within a practice, does this not imply a kind of linguistic idealism? Aucouturier accepts that it makes no sense to try to ground or justify the meaning of a rule outside of its application. This then leads to the questions of idealism: the capacity of language to reach outside itself and the possibility of incommensurability of systems of thought. Drawing on Anscombe, she argues that though the existence of rules does indeed depend on humans (a position she calls "partial idealism"), this neither prevents meaning from reaching reality nor rules out agreement in judgement between members of different linguistic communities.

One consequence of saying that language is a "practice" or "technique" is that *repetition* is essential to our use of words. For Wittgenstein, sensitivity to continuity between contexts seems to be part of the grammar of "being able to," both for language use and for other skill domains. In his "What is a Context?," Martin Gustafsson addresses the question whether such so-called "radical contextualism" implies an implausible fragmentation in our linguistic abilities that makes completely mysterious how we do in fact move intelligibly with language from one context to another. Gustafsson first lays out an argument put forth by Herman

Cappelen and Ernie Lepore that might be construed as undermining the kind of contextualism often ascribed to Wittgenstein. He then investigates the manner in which the concept of "context" is employed in such anti-contextualist arguments. In particular, Gustafsson finds a misleading, overly spatialized, metaphor of "inside/outside" that distorts our understanding of how participants in linguistic practices can typically move seamlessly in and out of new situations. Freed from the misleading metaphor, we can see that this fluidity is exactly what being "Master of a Technique" allows.

Counting Matters (Misak, Floyd, Agam-Segal, and Berts)

Wittgenstein is perhaps best known for writings on logic and language. Yet he spent an enormous amount of time and energy writing on the philosophy of mathematics as well, once commenting that his primary contribution to philosophy was in this area (cf. Monk 1990). But here too, his main focus was not on such questions as the "access problem," how finite material minds might have epistemological access to abstract, universal, necessary truths, but rather on shifting our perspective on such topics, giving them a "human face," as it were.

In "The Bolshevik Menace of Anthropological Mathematics: Ramsey and Wittgenstein," Cheryl Misak takes up the indispensable role played by Frank Ramsey for Wittgenstein's thought on mathematics in the crucial period after his return to philosophy and to Cambridge in 1929. Wittgenstein said (at some point between 1937–1944) that "mathematics is after all an anthropological phenomenon." This was an about-face from the view of mathematics expressed in the *Tractatus*, where he had said that mathematics consisted of "equations." Misak argues here that the shift was caused by Wittgenstein's discussions with Ramsey. Both came to see in 1929 that anthropological mathematics, or intuitionism, or what Ramsey called a "human logic," fit better with the philosophical picture they were starting to develop. Her argument is that by "anthropological," Wittgenstein meant "human," as Ramsey used the term.

The notion of practice has figured centrally in many depictions of the later Wittgenstein's vision of language. Yet, *Praxis* itself occurs comparatively infrequently in his writings. In her "*Technik* and *Praxis* in Later Wittgenstein," Juliet Floyd first notes the way in which "form of life" (*Lebensform*) assumed its place in Wittgenstein's thinking around 1936. Floyd goes on to show that *Praxis*, while pointing to the philosophical relevance of human activity, was too diffuse for Wittgenstein's purposes. He thus began to explore and deploy the more tractable notion of different "techniques" of use *within* a practice, first in 1937, in the context of his philosophy of mathematics. Floyd's discussion of Wittgenstein's appropriation of Spengler shows, moreover, that *Technik* reworks a key notion in Spengler's account of modernity, in particular his claim that mathematical achievements are signature expressions of a culture. Floyd thus brings out the significance of Wittgenstein's "anthropological" treatment of the foundations of mathematics in the period 1937–1944.

In "Mathematics and Aspect-Seeing," Reshef Agam-Segal examines the relevance of aspect-seeing to mathematics. Mathematics, he claims, often involves re-describing. Proofs, in particular, typically give us new ways of looking. This answers an old question: What is intellectually achieved by a mathematical proof? Agam-Segal shows how proofs can lead us to a sort of reflective position, akin to the mental position from which we see aspects—seeing things anew, as we can re-see  as a rabbit instead of a duck. From this perspective we recognize that our prelinguistic, "primitive," sense of things, contingent as it is, guides the formation of our practices and in some sense grounds them. Agam-Segal concludes by positioning his view between realism and antirealism in mathematics.

Kim-Erik Berts extends the consideration of proofs in his "Entering a Practice of Mathematical Proof: On the Difficulty of Teaching Proofs." He notes that within didactics research, there is widespread agreement on the importance of including proofs at all levels of mathematics education, a consensus reflected in many national curriculum documents. However, a common experience among educators is that proofs are difficult for students to understand, let alone produce. The main aim of Bert's chapter is to argue that research into the teaching of proofs can be furthered by incorporating the kind of focus on mathematical practice and context sensitivity that is found in Wittgenstein's philosophy. Although research in the teaching of mathematical proofs has already been strongly influenced by a focus on mathematical practice, Bert's argues that Wittgenstein's remarks on mathematical proofs can provide new insights into the issue of students' struggle to learn them. From this perspective, the rationale for teaching proofs can also be that the students need to learn an entirely new way of verifying a proposition. That students experience difficulties with proofs is seen as an expression of the struggle to enter this practice.

Agreeable Animals (Risjord, Rouse, and Cahill)

Wittgenstein wrote at *PI* §25 "Commanding, questioning, recounting, chatting, are as much a part of our natural history as walking, eating, drinking, playing." This clearly suggests that he believed our biological heritage had much relevance for our view of what are frequently taken to be quintessentially human activities. Yet, the relation between this heritage and these activities is much disputed, both quite generally and as it pertains to how we understand Wittgenstein's philosophy. This relation is especially acute when it comes to the question of how our natural history connects to the vexed question of normativity in human life: the fact that a strong notion of the concept "correct" pertains to so many of our activities, in ways that seem both applicable and not applicable to other animals. All three of the contributions of this section explore how such correctness can be understood as a social and contextual feature of natural human life.

In "Social Ontology, Evolution, and the Foundations of Practice Theory," Mark Risjord first notes several decades of work that seems to show that for joint action to be possible, sophisticated mental representations, such as propositional attitudes, mind-reading, mutual knowledge, and so on are required, and, therefore, that from the perspective of contemporary social ontology, there is simply no way that practices sufficiently sophisticated to support practice-theoretic accounts of representation could arise without the prior existence of humanlike representational capacities. Social accounts of representation in the vein of

Wittgenstein, Sellars, or Brandom, one might argue, are therefore simply non-starters. Risjord calls this the "social ontology objection" to practice-theoretic accounts of language. The object of his chapter is to rebut the social ontology objection by providing a minimalist account of joint action and thereby putting practice theory on a firm social-ontological foundation.

Wittgenstein's discussions of rule-following have often been taken as inspiration for practice-based social theories. Wittgenstein indicates a normative ordering and responsiveness expressed *in* what people do that is more fundamental than explicit interpretation *of* what they do. In "Practices, Normativity, and the Natural History of Human Biological Niche Construction," however, Joseph Rouse points out how theories of social practice have been inadequately attentive to Wittgenstein's occasional remarks on such activities as belonging to the "natural history of human beings." While eschewing any form of reductionism, Rouse argues for understanding the normative concerns constituted and sustained by ongoing practices as forms of biological niche construction, and thus as integral to the natural history of human ways of life in a perhaps even stronger sense than Wittgenstein intended. Normativity, Rouse argues, is a biological phenomenon and the diversity of normative concerns in people's practice-differentiated ways of life reflects an evolved, "two-dimensional" form of the one-dimensional biological normativity of most organismic lineages. This recognition has important implications for how normative concerns arise and interact, how their content is embedded in practices, and how those concerns are expressed discursively.

Although there might be disagreement as to precisely how much weight to give them, most commentators would agree that the notion of practice (*Praxis*) and related notions such as use (*Gebrauch*), custom (*Gepflogenheit*), and form of life (*Lebensform*) are important for understanding Wittgenstein's later thought. In "Avoiding Tac-on Theories of Culture," Kevin Cahill shows how regarding culture as entirely imposed from "above" leads to a failure to distinguish what Cahill describes as Wittgenstein's "anthropological naturalism" from a "sociological naturalism" of practice. This encourages philosophers to seek answers to problems in, say, the philosophy of mathematics by focusing most of their attention on the practices of working professional mathematicians. Whatever the merits of this trend in the philosophy of the sciences, in the case of reading Wittgenstein, too much focus on higher cognitive practices such as sophisticated theories, proofs, or even experimental activity risks bypassing what he is often targeting. For it is very far from obvious how his frequent discussions of extremely elementary linguistic and mathematical situations are supposed to shed light on the technical questions that are sometimes taken to be the very stuff of mathematics or logic. Drawing on some neglected considerations from Glifford Geertz on brain evolution and culture, Cahill shows how avoiding "tac-on" theories of culture clarifies key aspects of Wittgenstein's work.

References

Moi, Toril. 2017. *Revolution of the Ordinary: Literary Studies after Wittgenstein, Austin, and Cavell.* Chicago: University of Chicago Press.

Monk, Ray. 1990. *Ludwig Wittgenstein: The Duty of Genius.* New York: Free Press.

Wittgenstein, Ludwig. 1953/1997. *Philosophical Investigations.* Ed. G.E.M Anscombe and Rush Rhees. Oxford: Blackwell.

Zumhagen-Yekplé, Karen. 2020. *A Different Order of Difficulty: Literature after Wittgenstein.* Chicago: University of Chicago Press.

Part I

And End to Skimming

The (Hard) Practice of Reading: On the Use and Misuse of Wittgenstein for Literature

Nancy Yousef

I want to begin by making explicit my sense of this chapter as doubly displaced and in transit between two disciplines. My engagement with Wittgenstein has stubbornly persisted for many years in spite of the fact that in the field of literary criticism he has never attained the wide readership, let alone the dedicated following, of other twentieth-century philosophers upon whose work the elaborate edifice of "theory" has been constructed. Rarely invoking his name or quoting his words, I've nevertheless kept Wittgenstein close. He's a methodological muse, always urging thoughtful hesitation and reckoning with the specificity of contexts and occasions that evidently call for words and expressions of irreducible or untranslatable particularity. This is perhaps an overelaborate way of characterizing Wittgenstein's insistence (urgent and despairing insistence) that how we attend to "what we say when" matters (intellectually, ethically, emotionally); adapted to the relationship between reader and text it serves in the way a nagging conscience can serve, inhibiting all kinds of

N. Yousef (✉)
Rutgers University, New Brunswick, NJ, USA
e-mail: ny109@english.rutgers.edu

K. M. Cahill (ed.), *Wittgenstein on Practice*,
https://doi.org/10.1007/978-3-031-68656-6_2

liberties one might be tempted to take with a text, forbidding opportunistic or careless appropriations, urging generosity or at least patience in interpretive approach. Between the hermeneutic play of poststructuralism and the demystifying suspicions of cultural critique, Wittgenstein's concerns have been out of place.

But how do matters stand outside of literary studies, in the discipline of philosophy that used to be Wittgenstein's home? As intellectually vital as Wittgenstein scholarship has been in recent decades, it is also undeniably clear that his work has become marginal to the mainstream preoccupations of most North American departments of philosophy.[1] The highly ranked and well-regarded department at Rutgers, where I teach, is a prime example. It has many distinguished faculty members, and it undertakes many collaborative projects with cognitive science and computational linguistics. No one is reading or teaching Wittgenstein in that department. This neglect is symptomatic of—and perhaps even substantively connected to—an indifference to the history of philosophy more generally, notwithstanding exciting work in this area, especially in the history of analytic philosophy.[2] Wittgenstein is not being read at Rutgers, and neither is Nietzsche or Kierkegaard, neither is Locke, neither is Hegel—certainly not at the graduate level where the next generation of philosophy professors is being trained. What concerns me especially is the corresponding atrophying of reading as a philosophical practice, as a method of doing philosophy. Hence my interest in the practice of reading as Wittgenstein's work calls for it, and as philosophy calls for it. There is an important discussion to be had about the disappearance of classic texts, and of hermeneutic and exegetical practices in the discipline of

[1] On the eclipse of Wittgenstein, see, for example, Juliet Floyd, "Recent Themes in the History of Early Analytic Philosophy," *Journal of the History of Philosophy* 47:2 (2009) 188–198 and P.M.S. Hacker, "Analytical Philosophy: Beyond the Linguistic Turn and Back Again" in *The Analytic Turn*, ed. Michael Beaney (New York: Routledge, 2007) 125–141.

[2] Neglect of the history of philosophy and the relative decline of hermeneutic practices in the discipline have long been lamented. See, for example, Arthur Danto's 1984 Presidential Address to the APA, "Philosophy as/and/of Literature," *Proceedings of the American Philosophical Association* 58: 1 (1984) 5–20 and Bernard Williams, "What Might Philosophy Become?" (1997) in *Philosophy as a Humanistic Discipline*, ed. A.W. Moore (Princeton: Princeton University Press, 2009) 200–215. For an excellent sampling of recent work in the history of analytic philosophy, see *The Historical Turn in Analytic Philosophy*, ed. Erich Reck (New York: Palgrave Macmillan, 2013) and Floyd (cited above).

philosophy.[3] That topic lies outside the scope of this essay, though it is implicated in the questions of interdisciplinarity that I will be raising, and pertains to that sense of double displacement with which I began. If the reading of Wittgenstein is not taking place in departments of philosophy or in departments of literature, then neither do Wittgensteinean ways of reading and of thinking about language have a disciplinary home. Can we imagine co-creating a post-disciplinary space?

Wittgenstein and Literary Studies: Recent Appropriations

In the past decade, Wittgenstein has featured in some notable new work in literary studies. Some might see this trend as confirming his exile from Anglophone philosophy, literary studies being the final redoubt for thinkers who have become obsolete in the fields within which their work was once dominant (Freud is a good example). Literary studies of Wittgenstein have done important work towards integrating the philosopher into a richer account of European modernism, and this is to the benefit of both literary history and the history of philosophy. In particular, attention to the self-consciousness of artistic practice in modernism has been helpfully extended to understanding a related self-consciousness about the aspirations of philosophy in the same period, and the forms of argument and exposition most appropriate to those aspirations.[4] The inclusion of Wittgenstein in cultural and literary histories of modernism has substantively enriched those fields.

But Wittgenstein has also become a resource for literary theorists seeking an alternative to the ideological critique and "hermeneutics of

[3] I address this challenge in "Philosophy, Literature, and the Avoidance of Reading," *New Literary History*, 54: 4 (2024) 1459–1485.

[4] Marjorie Perloff, *Wittgenstein's Ladder: Poetic Language and the Strangeness of the Ordinary* (Chicago: University of Chicago Press, 1999); Michael Lemahieu and Karen Zumhagen-Yekplé, eds., *Wittgenstein and Modernism* (Chicago: University of Chicago Press, 2017); Karen Zumhagen-Yekplé, *A Different Order of Difficulty: Literature after Wittgenstein* (Chicago: University of Chicago Press, 2020); Greg Chase, *Wittgenstein and Modernist Fiction* (London: Anthem Press, 2022); Ben Ware, *Dialectic of the Ladder: Wittgenstein, the "Tractatus" and Modernism* (London: Bloomsbury, 2017).

suspicion" that have been dominant critical paradigms in the literary humanities for the past half century. Toril Moi's 2017 *Revolution of the Ordinary*—an explicit call for disciplinary reformation—has attracted the most attention, but Moi is by no means alone in turning to Wittgenstein as a resource for a "post-critical" approach to literary studies.[5] In the context of literary studies, the Wittgenstein of the *Philosophical Investigations* provides a kind of validation for various post-critical commitments, including the rejection of abstract jargon and systemizing theory, the foregrounding of readerly experience, the desire to give emotional attunement and ethical responsiveness a place in literary criticism. These are laudable goals, and insofar as Wittgenstein is adduced as a counterpoint to the usual theoretical touchstones, his name takes on the talismanic function of shielding post-critique from charges of naïveté or anti-intellectualism or quietism. Nevertheless, salutary polemical verve and clarity come at a cost. The recent appropriation of Wittgenstein in the service of reformist agendas in the field of literary theory may be a missed opportunity for more substantive interdisciplinary engagement at a moment when philosophy and literature need to make common cause in addressing the current crisis of the humanities. The steep drop in undergraduates majoring in liberal arts subject areas, the movement away from core curricular requirements of the "great books" genre, the dominance of cultural studies and historicism in literary studies, the increasing engagement with cognitive science and linguistics on the part of philosophy, the decline of foreign language study and corresponding disappearance of courses in foreign literatures—all these trends and developments of the past half century have had the unintended consequence of narrowing opportunities for students of philosophy and literature to come together as readers and as teachers of texts that have profoundly shaped the intellectual history of both fields.

My particular reservations about the use of Wittgenstein in some influential lines of "post-critique" has to do with the promotion and

[5] Toril Moi, *Revolution of the Ordinary: Literary Studies after Wittgenstein, Austin, and Cavell* (University of Chicago Press, 2017). See also Robert Chodat and John Gibson, eds. *Wittgenstein and Literary Studies* (Cambridge: Cambridge University Press, 2022); Elizabeth Anker and Rita Felski, ed. *Critique and Post-Critique* (Durham, NC: Duke University Press, 2017).

dissemination of an idea of "the ordinary" that assumes much of what is in question: its familiarity, its recognizability, its meaningfulness. I can describe the problem in this way: many students, on a first reading of the *Investigations*, will be quick to adopt an interpretation of "forms of life" or "language games" that settles on, or settles for, some notion of conventionalism, according to which Wittgenstein presents a vision of language as a shifting, culturally and historically relative landscape of concepts and rules. It takes time to understand that such an interpretation misses the complex challenges posed by Wittgenstein's examples—like a good answer that misunderstands the question. I see a comparable difficulty with notions of the ordinary and of ordinary language in some literary critical appropriations of Wittgenstein. The mistake, I think, lies in imagining that the region of the ordinary is just there, waiting to be noticed—as if it were something I could at any moment bring you to see by pointing down to the earth, say, as Raphael imagines Aristotle in the famous fresco of the *School of Athens*. The meaning and impact of that gesture depends on the larger complex picture in which it is embedded, on where and how Aristotle is pictured among the other thinkers populating Raphael's classical forum, and also on why and how the painting came to be made for *that* space, at *that* time.

In the context of the elaborate theoretical languages of contemporary critique, there may indeed be something valuable in the call to get back down to earth rhetorically and conceptually by committing to write in a way that acknowledges, and attempts to address the way that most people read and respond to literature. Nevertheless, Moi's urging that "critical reflection on literature and art should not be an elite activity…[but] should involve a dialogue between critics and ordinary readers" such that "the concerns of the latter should be taken seriously by the former" has to assume the givenness of the central term at issue (Moi 2017, 11). "The ordinary is something we all inhabit and participate in, often without even being aware of it," she asserts, while also rendering it fully accessible and conventional: "By the ordinary, I mean the everyday experiences, practices, and ways of living that are shared by most people, regardless of their social position or level of education" (Moi 2017, 7). Yet Wittgenstein's ordinary challenges precisely because it conjoins quotidian intimacies and familiarities with a confounding, obdurate, chronic *non*-awareness. His many remarks on overlooking the obvious offer a double provocation. First, the deflation of theoretical aspirations: "all explanation must disappear, and description alone must take its place" (*PI* §109). Secondly, and more crucially, there are the self-searching questions induced by the paradox: How do I know what I am missing if I miss it? How do I see what I overlook? Doesn't the fact of my overlooking it preclude my noticing it? Memorable remarks from the *Philosophical Investigations* emphasize disorientation ("A philosophical problem has the form: 'I don't know my way about," *PI* §123) and the paradoxical obscurity of the obvious: "The aspects of things that are most important to us are hidden because of their simplicity and familiarity. (One is unable to notice something—because it is always before one's eyes.)" (*PI* §129). My approach to the ordinary in Wittgenstein is an effort to grapple with the difficulty of recognizing it as such, and requires a commitment to an interdisciplinary practice of reading that Wittgenstein's writing itself seems to call for. I am less interested in using Wittgenstein to make an intervention in literary studies, therefore, than I am in establishing areas of collaboration between philosophers and literary scholars with a shared commitment to historically informed reading methods.

Reading Wittgenstein, in Practice

So how might one read Wittgenstein without assuming that the ordinary is just there—obviously, evidently, already familiar? I have found it helpful to begin by noticing how often the philosopher who is so committed to describing ordinary usage ventures into the territory belonging to aesthetics, to religion, and to their overlapping borderlands. Let's take, for example, the well-known remark first published in (what used to be called) "Part Two" the *Philosophical Investigations*: "the human body is the best picture of the human soul" (*PPF* iv, §25). Evocative and suggestive (hence attractive to any reader with a literary bent), the line also plays a substantive role in the series of remarks on responding to another's pain within which it appears. The line is one among several pivots in a sequence that works hard to displace epistemological questions so that ethical entailments can come into view.[6] The sequence begins with remarks on recognizing the pain of another:

> "I believe that he is suffering."–Do I also *believe* that he isn't an automaton? Only reluctantly could I use the word in both contexts.
>
> (Or is it like *this*: I believe that he is suffering, but am certain that he is not an automaton? Nonsense!) (*PPF* iv §19)

It's worth noting at the outset something obvious and obviously important: "The human body is the best picture of the human soul" has the structure of an assertion but—*and*—it depends on figurative meaning for its sense. The word "picture," anyway, as it is being used in the phrase, cannot be literal. This is to say that the famous aphorism belongs among innumerable remarks of Wittgenstein that make use of figurative language, and most readers of his work recognize that remarks of this kind are not mere embellishments or ornaments to the lines of thought being developed. Rather the pervasiveness of figurative language in Wittgenstein's prose needs—and has increasingly come—to be understood as belonging to his philosophical method, or practice.[7] When we recall the imperative articulated in *PI* §109—"All *explanation* must disappear, and description alone must take its place"—it is important to give some weight to the possibility that what the philosopher means by "description" includes forms of comparison, analogy, and figuration associated with poetic usage.

Can we understand poetic language as belonging to ordinary language? Is there even a sustainable distinction to be made between figurative and ordinary language?

Wittgenstein's unpublished notebooks and manuscripts contain many variations of the remark on the body picturing the soul. The earliest version I have found is "The face is the soul of the body." It appears among

[6] For this distinction in relation to *Philosophical Investigations*, see Stanley Cavell, *The Claim of Reason* (New York: Oxford University Press, 1979) 354–369.

[7] See, for example, David Stern, "Wittgenstein's Style and Texts" in Hans-Johann Glock and John Hyman, eds., *A Companion to Wittgenstein* (New York: Wiley-Blackwell, 2017) 41–55; James Conant, "Wittgenstein's Methods" in Oskari Kuusela and Marie McGinn, eds., *The Oxford Handbook of Wittgenstein* (New York: Oxford University Press, 2011) 620–646.

several remarks on the understanding of gestures in a pocket notebook of 1933, the contents of which are summarized on the flyleaf as: "Experience of [the] present moment is what's real / The concept 'perhaps' / Aesthetics / Meaning." Later versions of the formulation appear among sequences of remarks pertaining to pain language, the privacy of sensation, first- and third-person ascriptions of thought and feeling, and the contest between physiological and psychological explanations of behavior.[8]

Here is a partial list of occurrences:

"The face is the soul of the body" (MS 156a, 49r; *CV* 26)
"The human being is the best picture of the soul" (*RPPI* §281)
"Instead of 'attitude toward the soul,' one could say 'attitude toward a human being [*Mensch*]'" (*LWII* II, 38)

Variations of a remark reworked over the course of many years in varied contexts: Wittgenstein's writing practice demands a patient reading practice, and there's much to be said about the slightly different implications of these slightly different expressions. (Consider, for example, Wittgenstein's use of *Mensch* as a noun and of *menschliche* as an adjective modifying both body and soul.) Looking at these variants together, the undetermined and indeterminate status of the key term—"soul" (*Seele*)—begins to seem deliberately indeterminate, useful, and meaningful precisely because its signification hovers between literal and figurative possibilities. It's an extraordinary word that nevertheless belongs to ordinary language—and that very fact seems to have been of interest to Wittgenstein. "Much too little is made of it that we include the words 'soul' and 'spirit' in our educated vocabulary," he notes in the *Big Typescript* (*PO* §93).[9]

---

[8] For further discussion see Yousef, *The Aesthetic Commonplace* (New York: Oxford University Press, 2022) 124–135.

[9] A slightly earlier version of this observation appears also in the "Remarks on Frazer's *Golden Bough*" (1931). The source for both texts is MS 110, largely compiled in early 1931, and belonging to a set of manuscript volumes described as the "core record" of a "new stage" in his thinking in the years immediately following his return to Cambridge in 1929. See Joachim Schulte, "Text-Genetic Philosophical Note" for MSS 105–122, September 2016, Wittgenstein Archives at the University of Bergen. http://wittgensteinsource.org/Ms-110_m. Thanks to Kevin Cahill for directing me to the earlier version.

Wittgenstein's lectures of Easter term 1933 (contemporaneous with the *Big Typescript*) veered into wide-ranging discussions of religion, ethics, and aesthetics. G.E. Moore's voluminous notes memorializing the sessions include his account of the May 1 meeting during which Wittgenstein ventured to address meaningful usage of the word "soul":

> One great trouble our language gets us into is that we take a substantive to stand for a thing or substance...
> Cf. "soul" which has sometimes been described as something "gaseous." But others haven't meant by soul anything like this...
> If I restricted use of "soul" to such phrases as "his soul is at rest" or "his soul is easily stirred," you might say I'm denying that there is any soul: but you may mean by "Men have souls" simply that such propositions are true...
> By asking what he would say, & what he wouldn't, you can get at how he uses the word.
> Haeckel said "God is a gaseous vertebrate," meaning that that's what people meant.
> This is like saying "Soul is a gaseous human being" & answer is sometimes people so use this word, but sometimes not at all.
> James on "breath" and relation to soul.
> Being a substantive people try to find a substance for "soul"; e.g. it's really the brain, it's really the breath.
> A) "Soul is really activities of the body." B) "Can't you see that soul can't be body?"[10]

In the synoptic account that Moore published in *Mind* over twenty years later, he recalled Wittgenstein speaking more about aesthetics than about ethics during these lectures, especially in the context of addressing differences between philosophy and science—another recurrent theme of his writings during this period. According to Moore, Wittgenstein frequently insisted that certain forms of philosophical inquiry transpire "independent of any special results of science."[11] What did this entail as a matter of practice?

[10] Notes from May 1, 1933 in *Wittgenstein: Lectures, Cambridge 1930–1933, from the Notes of G.E. Moore*, eds. David Stern, Brian Rogers, and Gabriel Citron (Cambridge: Cambridge University Press, 2016) 318–320 (Moore's underlinings). Cited as *MWL* hereafter.

[11] G.E. Moore, "Wittgenstein's Lectures, 1930–1933," *Mind*, New Series 64: 253 (1955) 26.

In discussions of Frazer's anthropology, Darwin's theory of the emotions, and Freud's psychoanalysis, Wittgenstein identified their "wonderful representations," "descriptions," and "good similes" as specifically aesthetic modes "of representing a fact" (*Mind* 19–20). Moore's notes of May 26 and 29, 1933, record Wittgenstein saying, *à propos* of Freud: "all aesthetics is of [the] nature of giving a paraphrase, even if [the] same words also express a hypothesis. It is giving a good simile"; "what Freud says sounds as if it were science, but is in fact a wonderful representation … it's all excellent similes" (*MWL*, 356, 365). Arising in the context of wider discussion of the differences between science and philosophy, the remarks on the soul in the 1933 lectures suggest the entwining of conceptual and methodological questions about the future of philosophy with a particular attentiveness to the poetic inflections of simple terms that cannot be reduced to what is "empirically given." The words of the biologist Ernst Haeckel and the psychologist William James are adduced not as hypotheses to be rejected, but as "wonderful representations" calling for attentive interpretation. Here again are Wittgenstein's contrasting allusions:

> Haeckel said "God is a gaseous vertebrate," meaning that that's what people meant.
> This is like saying "Soul is a gaseous human being" & answer is sometimes people so use this word, but sometimes not at all.
> James on "breath" and relation to soul.
> Being a substantive people try to find a substance for "soul"; e.g. it's really the brain, it's really the breath.
> A) "Soul is really activities of the body." B) "Can't you see that soul can't be body?" (*MWL*, 20)

Haeckel's *Riddle of the Universe at the Close of the Nineteenth Century* (1900) is likely Wittgenstein's source, specifically a passage of speculative genealogy that appears in a chapter on the "Immortality of the Soul." "Much more widespread and still much respected is the view which ascribes a gaseous nature to the substance of the soul," Haeckel writes, "The comparison of human breath with the wind is a very old one; they were originally considered to be identical … After a time, this 'living breath' was identified as the 'vital force' and finally it came to be regarded as the soul itself."[12] Wittgenstein also points to William James—probably thinking of the 1904 essay "Does Consciousness Exist?" in which James is keen to demystify "consciousness" by proposing that what philosophers mean by "I think" is *really* "I breathe": "moving outwards, between the glottis and the nostrils," James writes, "[breath] is, I am persuaded, the essence out of which philosophers have constructed the entity known to them as consciousness." "Stream of thinking," therefore, "is only a careless name for what, when scrutinized, reveals itself to consist of the stream of my breathing."[13]

Is Haeckel's evocation of the "gaseous nature" of the soul just a "good simile," or an elucidation of the enduring association of soul with breath? Does James's demystification of "I think" dissolve the sense in which, as he himself recognizes, "breath" has ever been understood as the "original of spirit"? The incommensurability of these possibilities appears in the impasse at which Wittgenstein's remarks in the lecture abruptly end: Possibility A) Either one takes soul to be another, fanciful, word for what is *really* "activities of the body," or Possibility B) one resists the reduction to the material with an exasperated exclamation: "Can't you see that soul can't be body?"

Wittgenstein continues to work through the implications of this descriptive impasse in the subsequent term, as recorded in the *Blue Book*: "it may appear that ... we have worlds built of different materials," he observes, "a mental world and a physical world. The mental world in fact is liable to be imagined as gaseous, or rather, aethereal. But let me remind you of the queer role which the gaseous and the aethereal play in philosophy ... I mean, we already know the idea of 'aethereal objects' as a subterfuge, when we are embarrassed about the grammar of certain words, and when all we know is that they are not used as names for material objects. This is a hint as to how the problem of two materials, *mind* and *matter*, is going to dissolve" (*BB*, 47).

---

[12] Ernst Haeckel, *The Riddle of the Universe at the Close of the Nineteenth Century* (1900), trans. Joseph McCabe (New York: Harper and Row, 1905) 20.

[13] William James, "Does Consciousness Exist?" (1904) in *Essays in Radical Empiricism*, ed. Fredson Bowers (Cambridge: Harvard University Press, 1976) 19.

It is not unusual in Wittgenstein's writings of the 1930s and 1940s to find pursuit of questions about thought and mind arrested at a point where scientific or materialist explanation of psychological phenomena is starkly juxtaposed to a colloquial use of language, as happens in the 1933 lecture with the mutually exclusive positions A) and B). Irresolvable as the conflict may be (soul is just body/soul can't be body), it nevertheless does not fully account for the uses of "soul" that Wittgenstein himself had brought forward earlier in the lecture. "His soul is at rest"; "his soul is easily stirred": what the philosopher appears to be grappling with is the strange possibility that we commonly use the word "soul" without necessarily believing in its existence as a matter of empirical fact, but at the same time, it doesn't seem quite right to say that the word is merely a "careless name" (as James would have it).

The meaningful persistence of the word "soul" in ordinary language is a remarkable if easily overlooked matter of fact. I do not mean that its current usage is historically interesting, or that it survives only as a remnant of belief systems that have been superseded but not yet wholly extinguished. If I say, "he has a dark soul" I think you will understand me to mean that he is not a light-hearted type, or that he is troubled and in need of cheering, or in need of being left alone. What you understand me to mean will depend on the occasion of our conversation: am I reacting to a story you've told me? am I joking? am I warning you to beware? What neither of us needs in order to grasp the meaning of the phrase is an underlying belief in an immaterial substance animating the body. Thinking theoretically about the problem of mind and matter seems a distraction from thinking about the ready comprehension of Wittgenstein's idiomatic examples: "his soul is stirred" tells me something about him, or about the speaker's sense of him—something that really has nothing at all to do with the meaning the word "soul" would have if we were talking about theology or metaphysics or scientific psychology.

In this case as elsewhere, Wittgenstein's lectures of the early 1930s are notable not only for their forays into aesthetics and religion, but also for early indications of the philosopher's methodological appeals to the ordinary and the obvious. "Everything we want lies open to [the] inspection of everybody," Moore records him saying as early as 1930: "I am not trying to teach you new facts. I shall only tell you trivial things—what you all know" (*MWL*, 87). This very observation will sound familiar as an antecedent of memorable remarks in the *Investigations* (e.g., §89: "It is, rather, essential to our investigation that we do not seek to learn anything *new* by it. We want to *understand* something that is already in plain view"). Nevertheless, granting that the word "soul" has colloquial—even trivial—uses, why does the philosopher attend to it, and even insistently draw our attention to it? After all, to acknowledge that we habitually do make sense of soul seems to involve admitting something outlandish and vaguely mystical into the region of ordinary language. Isn't this just what Wittgenstein aims to avoid by urging us to "stick to the subjects of our everyday thinking" (*PI* §106)?

Strange as it must seem then, purposeful usage of the word soul by Wittgenstein, and his attention to its meaningful usage, needs to be understood as informing what ordinary language includes, and as belonging to regions of the ordinary that are habitually overlooked.

In a passage quoted earlier from the *Blue Book*, Wittgenstein speaks of our being "embarrassed about the grammar of certain words." Soul is, I suppose, one of those words. Does it help us to overcome our embarrassment to say that we only mean it figuratively, as a figure of speech? So, when I say, "he has a dark soul," do I *really* just mean "he's depressed" or anxious or sullen? Those are different descriptions—and perhaps more appropriate in some contexts (say, a clinical setting). But in other contexts those alternatives are not necessarily better, nor more accurate—insofar as accuracy is a matter of finding words to fit an occasion. "He has a dark soul": there are occasions when I am not sure it would make sense to say that my usage of the word is merely or only figurative.

Let me return to the well-known remark: "the human body is the best picture of the human soul." The sequence leading up to that remark involves several reorienting movements or pivots that serve to shift away from (the temptation of) speaking as if my relation to the suffering of another were a matter of believing or ascertaining something, to the unavoidable recognition (implicitly ethical, anyway non-epistemological) Wittgenstein terms holding an "attitude" (*eine Einstellung*) towards another. It is worth pausing on this surprising and confounding passage in the middle of the sequence:

> Religion teaches that the soul can exist when the body has disintegrated. Now do I understand what it teaches? –Of course I understand it –I can imagine various things in connection with it. After all, pictures of these things have even been painted. And why should such a picture be only an imperfect rendering of the idea expressed? Why should it not do the *same* service as the spoken teaching? And it is the service that counts. (*PPF*, iv §23)

Without speculating as to why Wittgenstein chooses to include this remark about religion teaching that the soul can exist when the body has disintegrated, we can at least notice (first) that he moves quickly to answer the question he anticipates—"do I understand what it teaches?"—with an emphatic "Of course I understand it," and (second) that the kind of understanding at issue here is evidently linked to imagining things, to recalling unspecified pictures that have been painted, to bearing in mind the meaningfulness of picturing.[14] Here is the ending of the sequence, immediately following "The human body is the best picture of the human soul" as a kind of development or elaboration.

> But how about an expression [*Ausdruck*] like this: "When you said that, I understood it in my heart"? In saying which, one points at one's heart. And doesn't one *mean* this gesture? Of course one means it. Or is one aware of using a *mere* picture? Certainly not. –It is not a picture that we choose, not a simile [*Gleichnis*], yet it is a picture-like expression [*bildlicher Ausdruck*]. (*PPF* iv, §26)

The phrase "When you said that, I understood it in my heart" and its accompanying gesture, "one points at one's heart," remind us of the possibility of earnestly meaning and believing what one cannot literally mean and believe. Some occasions call for, or will call forth, such emphatic, even passionate, expressions. If I acknowledge the possibility of using such expressions meaningfully ("Of course one means it" recalls "Of course I understand it"), then I commit myself to responsibility for the use of these words—choosing them, withholding them, and also spurning them. I avoid the responsibility if I get hung up on the triviality of the figure, or take it as an affectation, or quickly turn it into something else.

[14] For further discussion of the entanglement of aesthetic and ethical implications here, see *The Aesthetic Commonplace*, 140–147.

Doesn't "I understood it in my heart" *really* just mean "I empathize"? Does "I empathize" perform the same service as "I understood it in my heart"? Does embarrassment compel such evasions? Or does embarrassment at the meaningful use of such a phrase tell us something about pressures and appeals for response that belong to the language we use every day?

"In my heart" is indeed a trite image for how I understand what you meant on an occasion when you really need me to understand what you have said. But as the response called for in the moment, "it is not a picture that we choose, not a simile." I would add that it is not *not* a simile either, which is perhaps what Wittgenstein himself means in calling it a "picture-like expression." Addressed to another and understood by another, the affective import of the words and the gesture—their together forming a response—plainly matter, and in that sense the words are not a "*mere* picture."

"I understood it in my heart" belongs among an array of expressions that locate an emotion, a thought, a mood, a state of mind in the body. In the sequence I've been discussing, it is related to the "picture" of "thoughts in the head." Situated between these two pictures—of understanding in the heart and thoughts in the head—is the passage about what can now be seen as a related picture—that of the soul incarnate. The idea expressed by the soul in the body is no more or less out of the ordinary than familiar expressions linking feelings and dispositions to parts of the body—the heart's desire with the soul's conviction, for example. The kinship among such expressions is perhaps most explicit in this remark from late in the *Investigations*:

> "In my heart I've decided it." And one is even inclined to point to one's breast as one says it. Psychologically, this way of speaking should be taken seriously. Why should it be taken less seriously than the statement that faith is a state of the soul? (Luther: "faith is under the left nipple.") (*PI* §589)[15]

[15] Similar reflections immediately precede the aphorism "The human being is the best picture of the human soul" in *RPPI*: "Is it not important that for me hope lives in the *breast*? ... Why does a human being believe that a thought comes into his head? Or, more correctly, he does not *believe* it; he lives it. For he clutches at his head; he shuts his eyes in order to be alone with himself in his head." (*RPPI* § 278; 281).

"In my heart" is not just a metaphor standing in for the "real" site of decision-making (a process in the brain, for example). Nor does its meaning necessarily hang on a strict distinction between literal and figurative senses. The ordinary words serve in a way that a poet or a theologian or a painter (or a philosopher who works with descriptions alone) is best able to bring to light. Certainly Luther's startling intrusion in this remark *explains* nothing, but it can be understood to thicken the description of "this way of speaking." "In my heart" has a long history of giving expression to a particular kind of decisive resolution (the kind made in spite of what reason dictates, or as a matter of faith, or in defiance of the facts). The reference to Luther makes clear that "remarks on the natural history of human beings" (*PI* §415) such as Wittgenstein claims to be assembling necessarily include the accretions of cultural history as well. "Faith is under the left nipple": the phrase deploys literalism as itself a kind of trope, pointing to the visible body part that (literally) covers the heart where faith is (figuratively) inscribed.[16] The expression shuttles between figurative and literal meanings, and obviates the distinction between them.

"It is always pernicious to *forget* the picture & still use the same language," Wittgenstein warns, adding that "It is important to insist on the picture: as in psychoanalysis, we must make the picture conscious: *then* we can use the language without danger" (*MWL*, 87). The observation occurs in the same lecture as Wittgenstein's remark about endeavoring to teach "trivial things—things which we all know already." How is importance made manifest? Freud would say that repetition is a sure indication—instead of passing over, one is not able to let go—returns again and again. Recollection is at work here as well, for if it is pernicious to forget the picture, then that danger is averted by remembering, or being

[16] Two possible sources from Luther's sermons have been identified by Eike von Savigny (German editor of the *Investigations*), but Wittgenstein's rare references to Luther tend to be gestural and second-hand. For example: "One says God speaks to me in my *heart*: one speaks of the eyes, the hand, the mouth of God, but not the other parts of the body: Learn from this grammar of the 'God'! (I read somewhere, Luther had written that theology is the 'grammar of the word God')" (23 February 1937; *PPO* 211). Even so, the connection of Luther's name with a faith located in the heart suggests the cultural diffusion and inheritance of that association: Cf. Luther: "Whether I want to or not, when I hear the word Christ, there delineates itself in my heart the picture of a man who hangs on the cross." Qtd. in Joseph Koerner, *The Reformation of the Image* (University of Chicago Press, 2004) 160; on Lutheran iconography of the heart, 213–221.

reminded, or perhaps—what may be the same thing—becoming mindful of the picture. Wittgenstein's remark about the body being the best picture of the soul is one that I returned to again and again (as holding a problem and a resolution in so perfect a tension as to be almost, but not quite, a point of repose). I found in my own returns to this remark that it was one to which the writer also frequently returned. Its many variations, situated among reflections on different topics in the philosophy of psychology, has given me a better sense of how Wittgenstein's attention to talk about the heart and the soul discloses a temptation to fall back upon a hard distinction between literal and figurative usages and, in so doing, to avoid, or at least overlook, the ordinary—but by no means simple—entwining of aesthetic resources and ethical implications in the language we use with one another every day.

The ordinary in Wittgenstein can be surprising, embarrassing, discomfiting. Recognizing how a familiar combination of words serves what he calls our "real need" is always an arduous endeavor for Wittgenstein. Literary theorists who appeal to Wittgenstein for a model of reading that makes attunement, responsiveness, and even ethical insight seem like readily achievable tasks need to take the philosopher's own practice more to heart.

My aim here, however, has not (only) been to caution against appropriations of Wittgenstein in my own discipline that, in simplifying the ordinary, risk overlooking the utmost seriousness with which he attends to the aesthetic aspects of language. I hope also to provide a sense of how attention to the aesthetics of ordinary language, and especially to the figurative usages that are central to the practice of literary interpretation, may have something to contribute to philosophical engagement with Wittgenstein's work.

One final quotation:

> Is it misleading to speak of man's soul, or his spirit? So little misleading that it is quite intelligible if I say, "My soul is tired, not just my mind." But don't you say that everything that can be expressed by means of the word "soul" can also be expressed somehow by means of words for the corporeal? I do not say that … For the words, and also what we point to in explaining them, are nothing but instruments, and everything depends on their use. (*RPPI* § 586)

Philosophers and literary scholars need to work together in order to understand the "everything" that depends on our use of words. But first we will need to create a space between our disciplines where reading Wittgenstein, and reading in the ways Wittgenstein's work calls for, might have a place.

References

Anker, Elizabeth, and Rita Felski, eds. 2017. *Critique and Post-Critique*. Durham, NC: Duke University Press.

Cavell, Stanley. 1979. *The Claim of Reason*. New York: Oxford University Press.

Chase, Greg. 2022. *Wittgenstein and Modernist Fiction*. London: Anthem Press.

Chodat, Robert, and John Gibson, eds. 2022. *Wittgenstein and Literary Studies*. Cambridge: Cambridge University Press.

Conant, James. 2011. Wittgenstein's Methods. In *The Oxford Handbook of Wittgenstein*, ed. Oskari Kuusela and Marie McGinn, 620–646. New York: Oxford University Press.

Danto, Arthur. 1984. Philosophy as/and/of Literature. *Proceedings of the American Philosophical Association* 58 (1): 5–20.

Floyd, Juliet. 2009. Recent Themes in the History of Early Analytic Philosophy. *Journal of the History of Philosophy* 47 (2): 157–200.

Hacker, P.M.S. 2007. Analytical Philosophy: Beyond the Linguistic Turn and Back Again. In *The Analytic Turn*, ed. Michael Beaney, 125–141. New York: Routledge.

Haeckel, Ernst. 1905. *The Riddle of the Universe at the Close of the Nineteenth Century*. Trans. Joseph McCabe. New York: Harper and Row.

James, William. 1976. Does Consciousness Exist? In *Essays in Radical Empiricism*, ed. Fredson Bowers. Cambridge: Harvard University Press.

Koerner, Joseph. 2004. *The Reformation of the Image*. Chicago: University of Chicago Press.

Lemahieu, Michael, and Karen Zumhagen-Yekplé, eds. 2017. *Wittgenstein and Modernism*. Chicago: University of Chicago Press.

Moi, Toril. 2017. *Revolution of the Ordinary: Literary Studies After Wittgenstein, Austin, and Cavell*. Chicago: University of Chicago Press.

Moore, G.E. 1955. Wittgenstein's Lectures, 1930–1933. *Mind* 64 (253): 1–27.

———. 2016. *Wittgenstein: Lectures, Cambridge 1930–1933, from the Notes of G.E. Moore*. Ed. David Stern, Brian Rogers, and Gabriel Citron. Cambridge: Cambridge University Press.

Perloff, Marjorie. 1999. *Wittgenstein's Ladder: Poetic Language and the Strangeness of the Ordinary*. Chicago: University of Chicago Press.

Reck, Erich, ed. 2013. *The Historical Turn in Analytic Philosophy*. New York: Palgrave Macmillan.

Schulte, Joachim. 2016. Text-Genetic Philosophical Note, MSS 105–122, September. Wittgenstein Archives at the University of Bergen. http://wittgensteinsource.org/Ms-110_m.

Stern, David. 2017. Wittgenstein's Style and Texts. In *A Companion to Wittgenstein*, ed. Hans-Johann Glock and John Hyman, 41–55. New York: Wiley-Blackwell.

Ware, Ben. 2017. *Dialectic of the Ladder: Wittgenstein, the "Tractatus" and Modernism*. London: Bloomsbury.

Williams, Bernard. 2009. What Might Philosophy Become? In *Philosophy as a Humanistic Discipline*, ed. A.W. Moore, 200–215. Princeton: Princeton University Press.

Wittgenstein, Ludwig. 2009. *Philosophical Investigations*. Trans. G.E.M. Anscombe, P.M.S. Hacker, and Joachim Schulte, Revised 4th ed. New York: Wiley-Blackwell.

Yousef, Nancy. 2022. *The Aesthetic Commonplace*. New York: Oxford University Press.

———. 2023. Philosophy, Literature, and the Avoidance of Reading. *New Literary History* 54: 4 (2024) 1459–1485.

Zumhagen-Yekplé, Karen. 2020. *A Different Order of Difficulty: Literature After Wittgenstein*. Chicago: University of Chicago Press.

Part II

Learning to Go on

"Following a Rule" Is a Practice

Lars Hertzberg

A Customary Way of Acting

Consider two actions:

1. I see a sign with the word "EXIT" and beside it an arrow pointing to the right, I turn to the right and walk on.
2. A dog bares its teeth at me and I pull back.

Both responses are intelligible. The first is in accordance with a practice, the second is not. Pulling back may be "instinctive", or it may be a response I have learnt from bitter experience, or have been taught by my parents.

My being said to follow the sign requires a background of people customarily acting in this way, pulling back from the dog does not. The

L. Hertzberg (✉)
Åbo Akademi University, Helsinki, Finland
e-mail: Lars.Hertzberg@abo.fi

K. M. Cahill (ed.), *Wittgenstein on Practice*,
https://doi.org/10.1007/978-3-031-68656-6_3

background is a *conceptual* requirement: it is because of this background that my turning right at the sign can be described as following the sign. The background may not only be involved in answering the question *what* I did; in a given case it may also enter into the question *why* I did what I did: I turned right at the sign because I was on my way to the railway station and that sign points the way to the station.

On the other hand, it *is* possible (if not likely) that there should only have been one occasion when someone pulled back from a snarling dog; even so the response could be intelligible. This difference is sometimes expressed by saying that following a sign is a convention, pulling back from a dog is not.

Actions like following signs are a paradigm in Wittgenstein's discussion of meaning and understanding. Two points appear to be central to his remarks about this topic. One is that it is only against the background of things done at other times by me and/or by other people that there may be a distinction between a right and a wrong way of acting on the sign. The right/wrong distinction is connected with the idea that there is an agreement in interaction between those who participate in the practice – an agreement that may be voiced but which will in many cases show itself simply in the ways we act together. The other point is that, though following the sign is something I do at a particular point in time, the application of this description to what I do is logically dependent on the existence of an ongoing practice.

The temptation to be resisted here is to think that what makes it true to say that in turning right I am following the sign can be exhaustively spelled out in terms of what happens at the time of acting. On the one hand, this might be thought to be a mental occurrence in which the sight of the sign produces the thought that I should take a right turn. However, no mental occurrence by itself could make my action a genuine case of following the sign. Of course I *may* have this thought but that by itself would not be what justifies saying that I followed the sign.

On the other hand, the occurrence might be assumed to be some causal, say, neurological process leading from the perception of the sign to the appropriate movement of my limbs. Again, various neurological linkups are sure to take place, but they would not be what makes what I do following the sign. If I was following the sign I can justify what I did by pointing to the sign; I may have no idea of what happens in my nervous system.

While customary ways of acting play an important part in Wittgenstein's discussion of meaning and understanding, we may note that the word "practice" (German *"Praxis"*) itself occurs only five times in the *Philosophical Investigations*, and in none of these occurrences is it foregrounded.[1]

The words "practice" and "rule" are of course closely connected in the *Philosophical Investigations*. I shall try to say something about the nature of that connection.

PI § 199, is a central expression of Wittgenstein's conception of rules and practices:

> Is what we call "following a rule" something that it would be possible for only one man, only once in a lifetime, to do? – And this is, of course, a gloss on the grammar of the expression "to follow a rule".
>
> It is not possible that there should have been only one occasion on which only one person followed a rule. It is not possible that there should have been only one occasion on which a report was made, an order given or understood; and so on. – To follow a rule, to make a report, to give an order, to play a game of chess are customs (usages, institutions).

In *PI* § 54 Wittgenstein says that, in some cases,

> we say that [a game] is played according to such-and-such rules because an observer can read these rules off from the way the game is played – like a natural law governing the play. — But how does the observer distinguish in this case between players' mistakes and correct play? – There are characteristic signs of it in the players' behaviour. Think of the behaviour characteristic of someone correcting a slip of the tongue. It would be possible to recognize that someone was doing so even without knowing his language.

[1] On three occasions the word occurs in the combination "practice of the use of language", "linguistic practice" and "practice of language" (§§ 7, 21, 51). Here, the sense of the word seems closely akin to that of "language-game". On one occasion, Wittgenstein talks of the playing of chess as a practice (§197), and on one occasion he says that "following a rule" is a practice (§202).

One difference between observing a game and observing some natural occurrence is that, if I take what I am observing to be a game, I will assume that a distinction is in play between moves that are accepted and those that are not accepted within the game; not everything that happens during the game is in accord with the rules. Wittgenstein suggests that, even if we are not familiar with the game, I may catch on to the distinction, say, through noticing that someone corrected herself the way we notice someone correcting a slip of the tongue even if we do not understand the language she is speaking.

This might be taken to imply that a person could uphold the right/wrong distinction on his own. On the other hand, in § 202 Wittgenstein writes:

> … to *think* one is following a rule is not to follow a rule. And that's why it is not possible to follow a rule "privately": otherwise thinking one was following a rule would be the same thing as following it.

But of course, analogously, for a person to *think* she is *not* following a rule is not the same as *not following* it: to correct what she *takes* to be a slip of the tongue is not the same thing as correcting a slip of the tongue. When we think of someone as correcting a slip of her tongue we think of her as attempting to stick to what is a correct way of talking within the linguistic practice in which she is participating. What she does is a correction if it involves an attempt to conform to a practice. But this means that self-correction as such does not give sense to the notion of private rule-following.

G. P. Baker and P. M. S. Hacker argue that a practice may be limited to one individual, provided it would be *possible* for that individual to share it with others. They write:

> Since mastery of a technique is exhibited in acts which satisfy the criteria of correctness internal to the technique …, this manifestly guarantees the possibility of distinguishing between a person's thinking that he is following a rule and his actually doing so. It is, however, tempting to construct an explanation for the objectivity by reference to the idea that techniques are inherently *social.* On such an interpretation, "correct" … would be defined or explained by reference to what most qualified members of a practice do from case to case. (Baker and Hacker 1985, 164)

They call this a "radically mistaken" account, on the grounds that

> any definition of "correct" in terms of consensus or statistical regularity treats correctness as externally related to the technique of applying a given rule, whereas stipulating criteria of correctness is an essential part of explaining any rule whatever. (Ibid)

They are right in arguing that for an observer, say, to define a standard of correctness by invoking social consensus or the majority view within a collective would be for him to impose an external perspective. This argument requires some sorting out, however. For one thing, a distinction needs to be made between consensus and statistical regularity. Consensus is a matter of how participants think and talk about what they are doing, say, what they call correct or incorrect, whereas statistical regularity concerns the way participants tend to act. Furthermore, we should note that to call something correct or incorrect may either be something that is done in exercising a practice, or something done in describing it. Now the important point to note is that the latter is parasitic on the former. It is only to the extent that an observer is able to detect corrective or self-corrective remarks or manoeuvres that she has a ground for distinguishing between correct and incorrect ways of carrying out the practice. (The fact that some action appears to be done differently, say, by a minority of participants, does not in itself provide a basis for calling it incorrect: on the one hand, the variation may simply be disregarded; on the other hand, what looks to an observer like two ways of doing the same thing may in fact be two different actions within the practice, or the action may belong to a different practice.)[2]

When one of the participants calls a certain way of proceeding correct or incorrect, however, he is exercising a habit he has acquired in learning the practice. He is not making a judgment *about* an existing consensus. If construed in that way, his judgment would involve a logical circle: to the

[2] If there is disagreement, the observer must report that disagreement if her account is to be complete: if she pronounces one side correct she has ceased to be an observer. It is only to the extent that there is consensus that she is justified in calling a single way of acting correct.—Here, I will not go into complications such as the fact that there may be different levels of consensus. Thus, there may an "official" version of the practice, which is not in complete accord with the way the practice is actually carried out.

degree to which there is a consensus, it consists precisely in participants agreeing in calling ways of proceeding correct or incorrect. Nor is he invoking a statistical regularity, since the question what regularities are relevant can only be answered from within the practice in which judgments of correct and incorrect are made. (Of course he may say "But that's (not) the way we always do it!" but that then is a plea for agreement; if the matter is debated by the participants that is not a dispute to be resolved by empirical investigation; in fact, there may be no resolution if the disputants stick to their views.)

Now while Baker and Hacker are right in rejecting the idea that an appeal to consensus or regularity would be imposing an external perspective on a practice, still a one-person practice would not be imaginable, since there is no space in which the right/wrong distinction could have a life.

What enables us to speak of an activity as a practice, then, is not that in all cases there is a unique answer to the question of what is correct and what is incorrect, but that there is a use for the right/wrong distinction in the interaction between participants. We could express this point by saying that a practice is not socially defined but is dialogue-dependent, the relevant dialogue varying with the type of practice in question and the role of agreement within that practice.

Cora Diamond makes a closely related point in her essay "Rules: Looking in the Right Place", in discussing the prejudices affecting the discussion about communal agreement and rule-following:

> I have argued that the disputes in the philosophical literature about whether Wittgenstein thought that communal agreement makes possible some essential feature of rule-following are confused, and that such confusion is fed by the abstracting of "agreement" from the life into which it is woven…
>
> Consider once more the business of there being correctness and incorrectness in the applying of a rule. What do the prejudices do to us here? We ask what makes it possible for there to be a difference between correct and incorrect; what we will not ask is what the use of this word "correct" looks like in our lives, or in imaginably different ones. (Diamond 1989, 33)

This, however, is the question we *ought* to be asking. I shall get back to the question later.

"Fixed Rules"

The use of the word "rule" in the *Philosophical Investigations* is evidently a carry-over from Wittgenstein's talk of calculus in the early 1930s. One central feature of a calculus is that the sense of the terms used is constituted by their role in an activity (the way, for instance, the meaning of the logical constants is constituted by their definition by truth-tables). The role of the terms is determined by the rule-formulations introduced in defining the calculus. The right/wrong distinction in a calculus is applied by appealing to those formulations. The formulations precede the practice. To think of a calculus is to think of a deliberate human creation.

In Wittgenstein's later work, talk of calculus was superseded by talk about practice and language-game. In § 81, Wittgenstein cautions against the assumption that in the use of words, considered as a practice, we follow *fixed* (*"feste"*) or *definite* (*"bestimmte"*) rules "as in a calculus". In the next paragraph but one, he sketches out what a situation in which there are no fixed rules might look like:

> 83. Doesn't the analogy between language and games throw light here? We can easily imagine people amusing themselves in a field by playing with a ball like this: starting various existing games, but playing several without finishing them, and in between throwing the ball aimlessly into the air, chasing one another with the ball, throwing it at one another for a joke, and so on. And now someone says: The whole time they are playing a ball-game and therefore are following definite rules at every throw.
>
> And is there not also the case where we play, and make up the rules as we go along? And even where we alter them – as we go along.

In this example, I take it, the way the right/wrong distinction plays out is shown in the actual playing, including the corrective responses. The playing and the rules, we might say, are two sides of the same coin. For another instance of a shifting activity we might think of informal human conversation. During a get-together of friends, say, the conversational genres may shift—say, from political argument to requests for advice to reminiscence to lamenting to joking to attempts at consolation or mutual reassurance—spontaneously, seamlessly, without explicit agreement, and yet the participants may stay attuned to one another throughout the interchange, each being clear about what is going on, what kinds of contribution may be appropriate and which are not.

Here it is important to keep in mind the distinction between talk about *rules* and talk about rule-*formulations*. Wittgenstein points to the distinction in § 54 (which was already quoted in part):

> Just think of the kinds of case where we say that a game is played according to a particular rule.
>
> The rule may be an aid in teaching the game. The learner is told it and given practice in applying it. – Or it is a tool of the game itself. *[So far, this is about rule-formulations.]*– Or a rule is employed neither in the teaching nor in the game itself; nor is it set down in a list of rules. One learns the game by watching how others play it. But we say that it is played according to such-and-such rules because an observer can read these rules off from the way the game is played…

As well as in § 82:

> What do I call "the rule according to which he proceeds"? – The hypothesis that satisfactorily describes his use of words, which we observe; or the rule which he looks up when he uses signs; or the one which he gives us in reply if we ask him what his rule is? – But what if observation does not clearly reveal any rule, and the question brings none to light? – For he did indeed give me an explanation when I asked him what he meant by "N", but he was prepared to withdraw this explanation and alter it. – So how am I to determine the rule according to which he is playing? He does not know it himself. – Or, more correctly: What is left for the expression "the rule according to which he proceeds" to say?[3]

Rule-formulations, Wittgenstein says, may be articulated both in teaching and in carrying out the practice. Here the articulation of the rule is meant to precede or to accompany the action in time. (For an example of the latter, we may think of the case of a football referee signalling what rule had been infringed in ordering a free kick.) On the other hand, a rule

[3] Even so, Wittgenstein does not always seem to keep the distinction clearly in focus.

may be formulated on the basis of observation or self-reflection; here the formulation follows the doing. In the latter cases, the rule-formulation does not have a role in the activity itself. Indeed, Wittgenstein suggests, in some cases even observation or reflection may not enable us to articulate any rule. We could imagine that the case of the shifting games described in § 83 would be one in which no discursive formulation can be found. found. Nevertheless, the participants may be able to call out one another on violations of the game.

I would suggest that there is a way of talking about "living by the rules" or "breaking the rules" (say, on how to conduct yourself in high society) which does not presuppose the possibility of identifying and articulating any exhaustive set of rules; rules talk here is not to be understood literally; the word "rules" is rather used as a kind of "mass noun", a way of indicating that an activity can be regarded under the aspect of the right/wrong distinction. We may also note that some of the examples of rules mentioned by Wittgenstein are ones which *could* not be discursively articulated: rules for colour words, rules for reading. It would not be possible to put into words the occasions on which it would be correct to describe the colour of an object as red, or how to pronounce the word "word" (anyway, to someone who did not have command of phonological vocabulary). In these cases, instruction is carried out, not by rule-formulations but by means of samples.

Talking about "following rules", roughly speaking, is talking about an activity in which there are things one is supposed to do and things one is not supposed to do. Differently put, the right/wrong distinction is basic to rules talk, not the other way round.[4]

We should guard against the temptation to think that rule-formulations are essential to a practice: that is, to think of them as instrumental in guiding the actions of the participants, as well as in allowing an observer to understand what the participants are doing, like a mental picture guiding our grasp of the proceedings. Such an idea would, of course, be at loggerheads with Wittgenstein's thoughts about language and action. It is

---

[4] For a critical discussion of the idea of language as a rule-governed activity, see Oswald Hanfling, "Does Language Need Rules?", in his *Wittgenstein and the Human Form of Life* (London: Routledge, 2002), 51–65.

only in the context of a linguistic practice that a rule-formulation can be said to determine a way of proceeding. *Ultimately*, it is the way the participants act and respond that shows what the rule stipulates, and not the other way round. What makes it easy to disregard this is the fact that our linguistic communication happens within a network of linguistic practices in which we are already immersed. Thus, when someone introduces us to a new game, he draws on ways of applying words to which we respond immediately in the context. In most cases it does not occur to us that in a different context these same instructions might be applied in some other—maybe a radically different—way.

Following a rule-formulation is itself a practice that may be regarded under the aspect of rule-following. These matters are Wittgenstein's concern in the oft-quoted *Philosophical Investigations* §§ 198–202. He sums up his discussion there by saying "That's why 'following a rule' is a practice"—note, he does not say: "a practice is to follow rules".

Games vs (Other) Practices

It may be useful to take a closer look at games as examples of practices. I am thinking here of standardized games, games with names, such as chess, or football, or croquet, or patience. Let's begin by noting that we would not ordinarily call, say, playing chess or football a practice. This may be connected with the fact that games possess a number of peculiarities when regarded as human forms of activity. Because of this, they are at one and the same time both perspicuous and misleading models of human practices.[5]

In talking about games, I shall be talking about the place they normally hold in *our* culture; in other cultures they may sometimes have religious or legal significance. In most cases games constitute limited wholes that are easy to survey, often involving a set of discursively formulated rules through which the participants in the game consciously

[5] For an illuminating discussion of, among other things, what it means to relate to an activity as a game, see R. W. Beardsmore, "Art and Family Resemblances", *Philosophical Investigations* 18 (1995), 199–215.

identify their actions. Many practices, on the other hand, do not have sharply defined limits. If there are rules, they may be fleeting, and there are often individual variations in the way people consider it correct to proceed within a practice. While games, in many cases, are deliberately instituted, many practices arise and change spontaneously. Unlike what is the case with a large number of practices, participation in a game is usually undertaken consciously and voluntarily. There are rules defining how football is to be played, but there is no rule *of football* stipulating that anyone *has* to play football in any given situation. (There may of course be a rule of the school that students have to play football at certain times.) On the other hand, many practices are, as it were, mandatory: when someone dies, for instance, there are not only customary ways of arranging a funeral, but it is an important custom *to* arrange a funeral. Maybe our reluctance to call games practices is connected with this fact. Games are a kind of extra, a bonus activity. For an adult in our culture, it is something one may or may not decide to include in one's life.

What ties the players to a game, then, is first and foremost their commitment to playing it. The importance of a game is bestowed on it by players themselves and by onlookers, it does not primarily derive from any external concerns—still, external concerns *may*, of course, come to play a role in connection with the game. This does not mean that anything is permitted as long as you abide by the rules of the game, even if certain things may be permitted in games that are not acceptable otherwise, such as tackling in football. It will sometimes be said that we play games for our enjoyment, but this, I think, only means that participation is voluntary and that games lack utilitarian purposes: a very competitive person, to whom games matter a great deal, may not really *enjoy* playing at all. In competitive games, the importance of individual moves simply lies in their contributing to the aim of winning the game; how winning is defined, on the other hand, is in a sense arbitrary. On the other hand, some practices, such as putting up and following road signs, are instrumental: their purpose is to enable travellers to find their way. In the case of many other practices, the activity is primarily judged by criteria that are internal to the practice (singing a song, greeting, carrying out a ritual), though their importance may be bound up with other concerns.

Competitive games, on their part, have an instrumentality that is internal to the game.

It could be argued that the clearly circumscribed character typical of games explains their usefulness as models of human practices, but at the same time that is precisely the feature that makes their use as models risky, in tempting us to overlook the variety and complexity of human practices. (This is true not least of the concept of a language-"game".)

The fact that standardized games are used as paradigms of customary ways of acting, I would argue, leads to an impoverished account of customs and rule-following in the *Philosophical Investigations*. First, customary ways of acting are here being conceived as a binary thing: either you follow the rules or you don't. Now, viewing practices under the right/wrong distinction is adequate for most games, as well as for activities such as reading, spelling, continuing number series, etc. It also fits the conception of language-games formulated in the introductory part of the *Philosophical Investigations*, such as the various builders' games. However, in other cases, customs may be many-dimensional. Peter Winch, in *The Idea of a Social Science*, contrasts rules such as those of grammar which are fairly strict and unequivocal (e.g. that it is wrong to follow a plural subject with a singular verb), with standards of style which may be followed to varying degrees and in a variety of ways (1958, 53–54). In fact, it could be argued that the difference between rules of grammar and rules of style is not just a difference between strict and loose. Learning rules of grammar is primarily learning to conform to a pattern. On the other hand, in writing one may choose between stylistic standards depending on the stance one wishes to convey, the atmosphere one wishes to create, etc. Standards of style, we may say, are geared to a *spectrum* of meanings—allowing for infinite gradations of nuance—whereas rules of grammar are, in the main, bound up with a yes/no space.

It might be said, then, that Wittgenstein's choice of examples produces a conventionalist bias in his thinking about language and action: by this I mean an account that emphasizes a pupil learning to *conform* to certain patterns of speech or behaviour, rather than learning to make use of customary ways of talking and acting in order to express herself and to navigate her human surroundings.

Second, in the *Philosophical Investigations* there is little discussion of what practices are *for*. Why do we conform to practices? In describing the teaching of language-games Wittgenstein uses the German word "*Abrichtung*" which could be translated as "drill" (although both Elizabeth Anscombe's original translation and the version amended by Peter Hacker and Joachim Schulte use the word "training"). The word brings to mind a more or less mechanical process, as when you train animals or recruits to respond to commands. Wittgenstein's point is that primitive language-games are not taught by *explanation* (cp. *PI* § 5). He may have thought that the reason we follow rules does not really matter for his concerns. Nevertheless, in acting in a customary way, we do not normally just respond mechanically to a cue (though it is true that some practices presuppose rote learning, as when we learn the number system or the alphabet). The way we describe a person's actions in participating in a practice will in many cases depend on the place of the practice in the person's life and the context in which she is acting. Her joining in the practice expresses her understanding of the situation. And what she takes herself to be doing depends on the kind of practice in which she is taking part.

Here, it may be wise to heed Cora Diamond's exhortation to ask "what the use of this word looks like in our lives, or in imaginably different ones". For instance, some practices—such as the use of road signs, the practice of driving on the right- or left-hand side of the road, queuing up for a bus— are instrumental. If asked why we conform we will probably point to the practical advantages of general conformity. In these cases, it is easy to see what might be the consequences of not conforming. It is also easy to imagine circumstances in which one might be justified or even obliged to deviate from the practice. At the other end of the spectrum there are ritual practices, such as a wedding ceremony, an inauguration, a baptism, or a funeral. In such cases, as was said before, the rules are internal to the practice. Someone's failing to abide by the rules may not have any important practical consequences (though it may have legal significance)—ultimately, a breach of the rules matters only to the extent to which participants care. In the case of many practices, such as a class reunion, a book-signing event, a pool party, conversation at dinner, there may be a mixture of both kinds of concern.

A Case Study: Greeting

For an illustration of the mixture of concerns that may be involved in a practice, I should like to take a look at a practice of such as greeting. (As far as I'm aware, this is a subject Wittgenstein never discussed.)

It is plausible to think that greeting is a practice which is present in every human culture, though the form it takes varies greatly both between cultures and within a single culture. Greetings might be said to ritualize the moment when two individuals mutually acknowledge one another. Sometimes doing so may have a practical purpose, but in most cases it is simply what we do. In a given situation, your greeting someone is not simply a question of whether you choose to or not. Being encultured into a practice of greeting has two sides: on the one hand, there is the question of the cases in which one is supposed to greet someone, and on the other hand the question of *how* one is to do it. In learning to greet, the child gradually grows into a practice characterized by a complex grammar. You are supposed to greet people you know when you encounter them (though what it means to "know someone" may not be much more than that it is a person you customarily greet; compare the expression "nodding acquaintance"). You also greet people you meet in certain places or situations even if you do not know them, and you greet people in starting an interaction with them. Mostly, you don't greet people repeatedly if you meet them several times in a short span of time, though it may be polite to exchange a quick glance. Some of these requirements are strict, others are fairly loose. Again, the *forms* of greeting will vary in rather complex ways depending on the relation between the parties and the circumstances in which they meet, from a nod or a bow to a phrase of greeting, a handshake, kisses, or a hug; sometimes one may even be expected to stop and engage in a piece of small-talk or light banter. Within limits, one may choose the form. An important feature of greetings is the expectation of mutuality. The person who goes first has some space, as it were, to set the tone of the greeting. For someone to fail to respond to a greeting, or to choose a more restrained form, will usually be taken as a slight.

The culture, we might say, provides a finely tuned repertoire of greetings adapted to specific relations and situations. It is something we try to master in growing up, for fear of seeming socially inept by being inattentive, or by coming across as overly aloof or as overly effusive. It is worth noting that the rules of greeting are rarely articulated, except perhaps in teaching children to greet, or in explaining the practice to a foreigner. At the same time, we may adjust the culturally provided repertoire in a variety of ways to the immediate situation. There is elbow room for personal expression, whether spontaneous or deliberate. Thus, our way of greeting someone may be more or less warm or restrained. We may deliberately withhold our greeting to show our disapproval of a person or a group of people; on the other hand, we may go out of our way to be attentive to individuals who are generally liable to be ignored.

There is a passage in Abdulrazak Gurnah's novel *By the Sea* which brings out some of the tensions which may be involved in the practice of greeting. Here, the two protagonists of the novel meet again after many years:

> So there I was, standing inches in front of the assassin, the man who after years of dispute and humiliation has dispossessed my father of our house and all its contents, and about whom I had heard endless stories of callous deceit and depravity and shameless greed. In the intense vicinity in which we all lived, I had had to learn never to look at Saleh Omar's face whenever I recognised his shape in the street, never to show that I had seen him. I was too young to know how to ignore someone. If I saw and recognised him I would be bound to greet him, because that was the courtesy we had been taught from the morning of our first ages. (2001, 101)

It is easy to recognize the tension here between the almost involuntary habits of politeness that have been inculcated in the protagonist, and the restraint required by the demands of honour.

Conclusion

A practice such as greeting differs in a number of ways from the builders' game or a game of chess. The right/wrong distinction looks quite different in the case of greeting than in those other practices. Besides, this practice contains a spectrum of meanings, not just right and wrong ways of acting. Other practices, again, may deviate from these in a number of ways. It was not Wittgenstein's concern to give an account of what different practices are like. Rather, the focus on practices, connected with the focus on rule-following and language-games, was a way of drawing attention to the way the meaning of an action performed or of words uttered is dependent on the context of acting or speaking.[6]

References

Baker, G.P., and P.M.S. Hacker. 1985. *Wittgenstein: Rules, Grammar and Necessity*. Oxford: Basil Blackwell.

Beardsmore, R.W. 1995. Art and Family Resemblances. *Philosophical Investigations* 18: 199–215.

Diamond, Cora. 1989. Rules: Looking in the Right Place. In *Wittgenstein: Attention to Particulars*, ed. D.Z. Phillips and Peter Winch, 12–34. Basingstoke: Macmillan.

Gurnah, Abdulrazak. 2001. *By the Sea*. London: Bloomsbury Publishing.

Hanfling, Oswald. 2002. Does Language Need Rules? In *Wittgenstein and the Human Form of Life*, 51–65. London: Routledge.

Winch, Peter. 1958. *The Idea of a Social Science and Its Relation to Philosophy*. London: Routledge and Kegan Paul.

Wittgenstein, Ludwig. 2009. *Philosophical Investigations*, 4th revised ed. Translated by G. E. M. Anscombe, Peter Hacker and Joachim Schulte. Oxford: Wiley-Blackwell.

---

[6] I wish to thank Kevin Cahill and David Cockburn for helpful comments.

Kripke and Wittgenstein on Rule-Following: The Problem of Empty Philosophical Explanations

Oskari Kuusela

This chapter discusses the question, what kind of explanation is given when rule-following is characterized as a practice or as involving or based on communal agreement, and how Wittgenstein's account of rule-following should be understood in light of his philosophical methodology. I argue that certain kinds of explanations, often attributed to Wittgenstein and discussed in this chapter with reference to Kripke, that treat communal agreement as a condition of possibility and a ground of rule-following, are problematic. Such explanations are not consistent with Wittgenstein's philosophical methodology, and closer inspection reveals them to be empty pseudo-explanations that cannot do the intended philosophical work. Thus, as I argue, Kripke's account merely pushes the problem about rules one step further, where it arises again as a problem about understanding communal agreement. Instead the characterization of rule-following as a practice is better construed as clarificatory

O. Kuusela (✉)
University of East Anglia, Norwich, UK
e-mail: O.Kuusela@uea.ac.uk

K. M. Cahill (ed.), *Wittgenstein on Practice*,
https://doi.org/10.1007/978-3-031-68656-6_4

description that ascribes a role to linguistic practices and communal agreement as the background or context against which instances of rule-following, having certain intentions and understanding meanings are possible.

Wittgenstein's Method

Before discussing rule-following, a few remarks are in order about Wittgenstein's methodology or philosophical approach. This is important exegetically because it provides conditions of adequacy for an interpretation of Wittgenstein's account of rule-following. In order to understand Wittgenstein's philosophical views on particular topics, one needs to pay attention to his views about philosophy and its methodology. Irrespective of merely exegetical considerations, taking into account Wittgenstein's method is also philosophically important in that what he says typically emerges as more rigorous and philosophically interesting, if we understand what he is trying to do. As I aim to show, his discussion of rule-following is no exception. In particular, taking into account Wittgenstein's rejection of philosophical theses or theories and certain kinds of philosophical explanations cannot be left merely as an afterthought, as if we could first work out what his view on X is and only subsequently how his view on X doesn't constitute a thesis, theory or a rejected kind of explanation. This is comparable to thinking that one could grasp what 'I will stick with you whatever happens' means, without understanding whether it is said as promise or a joke. How we understand the intended use or logical status of Wittgenstein's statements, in other words, affects their meaning and has direct consequences for grasping the philosophical point Wittgenstein is making. As will become evident, my understanding of what kind of explanation Wittgenstein is offering differs significantly from Kripke's. (See Kripke 1982, 69ff. for his rather quick comments on Wittgenstein's rejection of theses.)

Here are some relevant remarks on philosophy and philosophical methodology from the *Philosophical Investigations*:

> [...] we may not advance any kind of theory. There must not be anything hypothetical in our considerations. All *explanation* must disappear, and

> description alone must take its place. And this description gets its light – that is to say, its purpose – from the philosophical problems. These are, of course, not empirical problems; but they are solved through an insight into the workings of our language, and that in such a way that these workings are recognized – *despite* an urge to misunderstand them. The problems are solved, not by coming up with new discoveries, but by assembling what we have long been familiar with. Philosophy is a struggle against the bewitchment of our understanding by the resources of our language. (PI §109)

> Philosophy must not interfere in any way with the actual use of language; in the end it can only describe it.
> For it cannot justify it either. [...] (PI §124)

> Philosophy just puts everything before us, and neither explains nor deduces anything. [...] The name "philosophy" might also be given to what is possible before all discoveries and inventions. (PI §126)

> The work of the philosopher consists in marshalling recollections for a particular purpose. (PI §127)

> If someone were to advance *theses* in philosophy, it would never be possible to debate them, because everyone would agree with them. (PI §128)

> In philosophy no inferences are drawn. "But it must be like this!" is not a philosophical proposition. Philosophy only states what everyone concedes to it. (PI §599)

I eschew any long discussion of the interpretation of these remarks, and simply state how I understand them.[1] First, what Wittgenstein in general means by philosophical theses or theories is standard: they are assertions about exceptionless necessities or possibilities put forward as true/false propositions (cf. PI §§130–131). Regarding §109, I won't attempt any general characterization of philosophical explanations, as there are many kinds of them, and my focus will be on a specific type. However, the kind of hypothetical explanations that Wittgenstein rejects are exemplified by accounts that involve postulating hidden structures such as the logico-syntactical rules shared by every possible language that the *Tractatus* postulated, and in the case of rule-following peculiar mental

[1] See Kuusela (2008, 2019b) for more detailed discussions of Wittgenstein's method, including his rejection of philosophical these and theories, and the justification of the interpretation of his methodology I assume.

states and processes that can do what empirical processes and human minds in the usual sense cannot do, for example, run through an infinite series in a fraction of a second, or provide the right answer without failure, like a super rigid mechanism that cannot malfunction and never gets worn out unlike physical mechanisms (cf. PI §§188–189, 193–194).

Another relevant sense of explanation in the case of the rule-following, and the one I focus on below, is explanations that seek to establish the ground or foundation for something by stating necessary conditions of possibility for what is to be explained. Thus, one might try to explain rule-following or meaning by arguing that its possibility depends on such and such which constitutes a necessary condition for rule-following or linguistic meaning. Understood in this way, Wittgenstein's rule-following discussion would have the following form: he first rejects mentalistic explanations of rule-following where the meaning of the sign is fixed by a mental state or a process. Similarly, he rejects explanations of rule-following in terms of dispositions. So far this is uncontroversial, and I agree that Kripke's discussions of these critical points of Wittgenstein's are illuminating (see Kripke 1982, Chapter 2). But hereafter things get tricky. It seems that a natural question to ask next is: very well, and what *does* explain the possibility of rule-following and meaning a sign in a specific way? Here many Wittgenstein-interpreters, including Kripke, have identified customs or practices or the agreement of a community as Wittgenstein's answer (see references in footnote 8). On such accounts, the possibility of rule-following or meaning a sign in a certain way depends on there being a practice of rule-following that is upheld by the agreement of the community of rule-followers which then justifies following the rule in a certain way. But as I argue, Wittgenstein rejects this kind of explanations. The problem with them is described in the *Remarks on the Foundation of Mathematics*:

> To what extent can the function of a rule be described? Someone who is master of none, I can only train. But how can I explain the nature of a rule to myself?
>
> The difficult thing here is not, to dig down to the ground, but to recognize the ground that lies before us as the ground.

> For the ground keeps on giving us the illusory image of a greater depth, and when we seek to reach this, we keep on finding ourselves on the old level.
>
> Our disease is one of wanting to explain. (RFM VI §31)

I will come back to this later. For now, let us treat this and the preceding considerations as a reason to pay attention to what kind of explanations we want to ascribe to Wittgenstein or offer ourselves in developing Wittgenstein-inspired views independently of exegesis.

The four remarks I quoted (§§124, 126–128, 599) can be summed up as follows. Philosophy is in the business of clarifying something which we already know, and of which we therefore only need to be reminded of. Nothing hypothetical, no hidden entities, structures, states or processes need to be postulated for the purposes of clarification, but description suffices. (As I hope to make clear below, this point about description – which seems disagreeable and/or foreign to many analytic philosophers – is not an arbitrary limitation. It is motivated by problems with relevant kinds of philosophical explanations.) Moreover, regarding the notion of agreement in §§128 and 599, insofar as philosophy's task is to clarify, only what is agreed upon as achieving its clarificatory purpose counts as philosophy in the strict sense. The rest is attempts to clarify (or perhaps preparations for it), and as Wittgenstein explains in his lectures, if the clarifications don't achieve their purpose, one needs to put them to the side and try something else. 'I won't say anything that anyone can dispute. Or if anyone does dispute it, I will let that point drop and pass on to say something else' (LFM, 22).[2]

The point of this last remark can be outlined as follows. What can be employed to clarify philosophical matters depends on what exactly those to whom the clarifications are offered find problematic about a given issue and on their philosophical assumptions. Consequently, one may need to say different things to clarify a matter depending on whom one is talking to, what exactly they find problematic, and what is or is not controversial to them. (A clarification cannot help if it assumes something that the interlocutor rejects.) As the point can be summed up: just as one needs the right tools for any job, so for philosophical clarification.

---

[2] For a more detailed discussion of these remarks and Wittgenstein's notion of agreement in philosophy, see Kuusela (2008, Chapter 6.4).

Wittgenstein's grammatical statements then are intended as instruments of clarification, not theses. Like philosophical theses, statements of grammatical rules constitute expressions of necessity and possibility,[3] but in the context of Wittgenstein's philosophy they are employed as modes of describing or representing the objects of study rather than as true/false theses which the objects allegedly must correspond to. That is, as he emphasises, philosophical accounts constitute models to be used as objects of comparison with the help of which we clarify specific aspects or features or language use, but without claiming that those characterizations show what *all* relevant instances *must* be like (PI §§130–132). Unlike philosophical theses, Wittgensteinian clarificatory models can thus tolerate exceptions. Moreover, because they are not put forward as true/false theses, such clarifications don't exclude one another like true theses exclude any incompatible ones. Consequently, different models can be used to clarify different aspects of complex objects of study.[4] It is worth emphasizing, however, that even though Wittgensteinian clarifications don't constitute true/false theses, this doesn't imply that they cannot be assessed for correctness or are not truth-apt. It merely means that the truth of Wittgensteinian clarifications ought not to be understood on the model of theses. Rather than depending on correspondence with postulated non-empirical necessities in reality, the truth of Wittgensteinian clarifications is judged on the basis of the clarificatory work they can do. Truth then is not a property of such clarifications and not anything they contain like theses. Rather, truth is something that clarifications can bring to view and help to understand. Wittgensteinian clarifications, in other words, can help us understand how things are.[5]

I take the preceding points about clarifications to hold for Wittgenstein's remarks about rule-following too. Were this not the case, his work would suffer from internal tensions, but I see no indication of this. So much for the preliminaries; let us turn to rule-following.

[3] Articulating such rules is one of Wittgenstein's methods of clarification amongst several, including the method of using simple language-games and real or invented natural history as clarificatory models. See Kuusela (2019b), chapters 5 and 6 for discussion of these methods.

[4] See Kuusela (2019b) Chapters 6.5 and 6.6 for discussion.

[5] For the notion of truth in Wittgenstein, see Kuusela (2019b), Chapter 6.6. I discuss this in relation to Nietzsche and the questions he raises about the value of truth in philosophy in Kuusela (2023b).

Kripke on Wittgenstein and Rule-Following

As noted, I think Kripke has explained well Wittgenstein's critical arguments against conceptions of rule-following as a mental state, a mental process or a disposition to act.[6] These arguments explain how the problem about rule-following arises—although I would not agree with Kripke that Wittgenstein's purpose is to articulate scepticism about meaning, and therefore I don't agree with his attribution of a *sceptical* paradox to Wittgenstein (see section "A Different Picture: Wittgenstein on Rule-Following"). I won't discuss these negative arguments here, but focus on how Kripke construes Wittgenstein's solution to the problem or rule-following, and whether this account can explain how rule-following or meaning something specific by one's signs is possible. Kripke formulates the problem about rule-following, construed broadly as a problem about meaning and language, in the following ways: 'This, then, is the sceptical paradox. When I respond in one way rather than another to such a problem as '68+57', I can have no justification for one response rather than another [i.e. 125 by plus rather than 5 by quus]' (Kripke 1982, 21; square brackets by OK). And: '"How can we show any language at all (public, private, or what-have-you) to be possible?" [...] Wittgenstein's main problem is that it appears that he has shown all language, all concept formation, to be impossible, indeed unintelligible' (Kripke 1982, 62).[7]

The question I'm especially interested in in response to this way of formulating the problem about rule-following is whether Wittgenstein, moving on from his critical arguments, wants to give some kind of alternative explanation for the possibility, or the ground or foundation, of rule-following and linguistic meaning. With regard to this, Kripke

[6] *Pace* Kripke, it is important to distinguish action from behaviour (and associated dispositions), due to actions being intentional (cf. footnote 13 below).

[7] Another point of disagreement is that, unlike Kripke, I don't think that the problem about rules 'applies to all meaningful uses of language' (Kripke 1982, 7), because I don't think Wittgenstein has a theory of meaning as constituted by rules (see Kuusela 2008, Chapter 4). Rather, Wittgenstein recognizes a variety of different kinds of instances of meaningful language use and maintains that the concept of meaning is too complex to be captured with reference to anything common to all cases, such as rule-governedness. But this can be left to the side. The points about rules pertain to linguistic meaning to the extent that it does depend on rules, and I agree that rule-governed uses constitute an important class of the meaningful uses of language.

proposes that in place of mental acts or processes, or dispositions, Wittgenstein wants to establish communal agreement about linguistic meaning, or how to follow a rule, as what justifies following a rule in a certain way or understanding the meaning of a sign in some specific way. It may be worth noting that Kripke doesn't use the terms 'ground' or 'foundation'. I'm uncertain how significant this is, but perhaps talk about 'ground' or 'foundation' might suggest a straight solution as opposed to a sceptical one. (For Kripke's notion of a sceptical solution, see 1982, 66–67.) As far as I can see, this distinction won't affect any of my points, however. Kripke's basic idea is that the agreement of the linguistic community fixes what is the right way to follow a rule or what our signs mean.[8] Or as he puts it: 'Wittgenstein proposes a picture of language based, not on *truth conditions*, but on *assertability conditions* or *justification conditions* [...]' (Kripke 1982, 74). As I cannot see any important difference between claims about what language use is based on or how it is possible, and claims about such and such as the ground or foundation of language, for example that assertability or justification conditions constitute the ground or foundation for language use, I will keep using the notions of foundation, ground and grounding below.

But the points in section "Wittgenstein's Method" about theories, theses and explanations should make us pause. It certainly seems that Kripke and other similar readings attribute to Wittgenstein a thesis about the nature of rule-following, linguistic meaning or language use as grounded on communal agreement. In so doing they risk making Wittgenstein's philosophy contradictory, given Wittgenstein's rejection of theses and theories. But let us leave to the side as merely exegetical the question about Wittgenstein's consistency. Independently of that, I want to emphasize that it is problematic to assume that because Wittgenstein seeks to dismantle certain theses about the possibility of rule-following, meaning or language, he must be aiming to substitute for them some other thesis of similar type regarding their possibility, as if the question were merely what the real ground or foundation of rule-following, meaning or

[8] A similar view is held by, for example, Norman Malcolm (1986), John McDowell (1984), and Crispin Wright (2001), although McDowell later criticizes himself for this (McDowell 1992, 49). For discussion of Malcolm's interpretation, see Minar (1991, 218–221).

language is. This is to assume that the only way to do philosophy is to put forward theses. But we should remain open to the possibility of Wittgenstein doing something different, as he himself insists. This is to articulate, by means of reminders and clarifications, a perspicuous view of the concept of rule-following as well as those of language and meaning to the extent that meaningful language use is a rule-governed activity. Such a perspicuous view must then avoid falsifications that arise when philosophers try to force the phenomenon of rule-following, and together with it those of meaning and language, into some philosophical straitjacket or another, whether its mentalist, dispositionalist, communalist or something else.

Although this may at first sight seem hysterical—what could be wrong with philosophical explanations in terms of necessary conditions of possibility given their venerable history since Kant at least—I hope to make the case that this issue is worth careful consideration. I argue that there's a philosophical problem with the desire to find a general metaphysical justification—a philosophical ground or foundation—for rule-following, linguistic meaning or the possibility of language. Indeed, it is the desire for such general justification, which everyday phenomena don't seem able to satisfy, that creates the philosophical problem about rule-following in the first place. By contrast, the point of Wittgenstein's rule-following discussion, on the reading I propose, is that we can only come to see the phenomenon of rule following in the right light, once we have released ourselves from the desire to find a general justification for rule-following or the meaningfulness of linguistic signs or for language use. Part of Wittgenstein's purpose thus is to clarify how the search for a general justification or a ground or foundation is motivated by problematic philosophical ideas that stand in the way of a clear comprehension of what rule-following is, as we know this phenomenon from everyday life.[9]

[9] By talking about releasing ourselves from a certain kind of philosophical desire, I don't want to suggest that Wittgenstein is engaged in philosophical therapy, whereby he merely wishes to show that our philosophical questions are confused without offering any positive accounts instead. Arguably, such therapeutic interpretations misconstrue Wittgenstein's philosophical approach and the notion of dissolution in his philosophy, tacitly assuming that the only way to say something positive in philosophy is to put forward theses or theories. For criticisms of therapeutic interpretations, see Kuusela (2019a, 2019b, 39–40, 174–175, 2023a).

A Different Picture: Wittgenstein on Rule-Following

There's no doubt that Wittgenstein describes rule-following as a practice (a custom or an institution; PI §199), and that—in some sense—he is suggesting that this observation can help to dissolve the paradox that arises if rule-following is understood as a matter of interpreting the sign as having a certain meaning and fixing its meaning in this way. The paradox, in short, is that if we can bring any action into accordance with the rule by suitably interpreting it, then there's no accord or disaccord (PI §201). That is, if we try to account for rule-following in terms of how rule-followers interpret linguistic signs, talk about accord or disaccord loses its sense. Every interpretation of a sign is correct, but because correct is no longer contrasted with incorrect, nothing is correct as opposed to incorrect. A sign means whatever one interprets or takes it to mean.

Earlier in the 1930s Wittgenstein discussed this problem with reference to acts of meaning and intention: 'Every sign is capable of interpretation; but the meaning mustn't be capable of interpretation. It is the last interpretation' (BB, 34), whereby 'meaning' refers to 'a process accompanying the saying'. And: 'The intention seems to interpret, to give the final interpretation' (PG, 23). On such a view (which he rejects) any sign, including mental images and mathematical formulae in one's mind, is open to interpretation, but meaning or intending a sign in a certain way is not. Rather, meaning or intending is the last interpretation that fixes the meaning of the sign. We have already seen the implications of this: if a sign means whatever one means by it or takes it to mean, there's no such thing as understanding a sign correctly or incorrectly. Consequently, something like Kripkean meaning-scepticism seems to follow—except that there's something very peculiar about Kripke's scepticism about meaning that I want to bring out. For Kripke's suggestion that there's a problem of meaning-scepticism, such as his book articulates, evidently assumes that somehow he has managed to spell out this problem in his book. But in order for this to be possible it must be possible to talk meaningfully. This explains the paradoxical air surrounding Kripke's meaning-scepticism: if there's no meaning, how did Kripke manage to articulate the sceptical problem in the first place, and how did we manage to understand him? Somewhere something has gone wrong. In the *Investigations*, Wittgenstein singles out this feature of the philosophical situation as indicating that we are in a muddle:

> That there is a misunderstanding here is shown by the mere fact that in this chain of reasoning we place one interpretation behind another, as if each one contented us at least for a moment, until we thought of yet another lying behind it. For *what we thereby show* is that there is a way of grasping a rule which is *not* an interpretation, but which, from case to case of application, is exhibited in what we call "following the rule" and "going against it". (PI §201; first italics by OK)

In other words, somehow we manage to successfully talk about interpreting our signs and giving them meaning by interpretating them, until we conclude from problems with this account that it cannot explain meaning—and that therefore there's no meaning! But as I hope is now clear, this conclusion stands on shaky grounds, because what it allegedly denies, i.e. that our words have definite meanings, was assumed and relied upon in reaching the very conclusion that words have no definite meanings.[10] Kripkean scepticism about meaning, therefore, is a muddle that arises from paying insufficient attention to how the conclusion about meaning-scepticism was reached in the first place. So much for the problem as Wittgenstein himself describes it as opposed to Kripke. What about the solution? How do we find a way out of this muddle?[11]

I have urged caution about rushing ahead without paying attention to our philosophical instincts and habits, and against taking the notion of a practice as something that grounds the possibility of rule-following or

[10] For a similar account of the problem of the paradox of interpretations explained slightly differently, see Minar (1991, 214, 1994, 58–61).

[11] More precisely, Wittgenstein's solution is a dissolution in the sense that, rather than answering the problem positively in the terms in which it was articulated, thus accepting the problem as genuine one, the goal is to make the problem disappear by changing the way we think about linguistic meaning. Here, contrary to therapeutic readings (as well as McDowell 1992, 47), introducing a different view is an essential part of dissolution. For an account of Wittgenstein's notion of dissolution, see Kuusela (2023a, 2023b). I'm unsure whether this account would be acceptable to McDowell as a path between quietism and constructive philosophy, but it can be seen as offering one (see 1992, 51).

meaning something specific by signs. To emphasize, however, the reason for this is not that Wittgenstein rejects theses, theories or explanations. Although it is exegetically important not to ascribe contradictory views to him insofar as there are alternatives, I want to focus on the philosophical or systematic side of the issue. But then, what is the problem with saying that the agreement of the community fixes the meaning of the rule or the sign or that communal agreement justifies taking the rule or sign in a certain way (e.g. that 68 + 57 = 125 by plus, not 68 + 57 = 5 by quus)? In order for us to be able to appreciate Wittgenstein's dissolution of the problem of rule-following, this apparent explanation must first be cleared out of the way.[12]

Ultimately the problem is that this kind of explanation gets us no further, just as Wittgenstein says in the quote in section "Wittgenstein's Method" from the *Foundations*: 'For the ground keeps on giving us the illusory image of a greater depth, and when we seek to reach this, we keep on finding ourselves on the old level.' How does this problem of finding ourselves on the old level arise in connection with Kripke's attempt to solve the rule-following problem in terms of communal agreement? As I argue next, the same problem that originally arose about rule-following arises again about the practice or communal agreement, when conceived as the basis of or ground for rule-following that determines and/or justifies how a rule is to be followed. If this is correct, there can be no general metaphysical justification for following rules of the kind that Kripke imagines can be given in terms of communal agreement.

First, in order for the practice or the agreement of community to constitute a standard with reference to which certain uses of signs (as opposed to others) can be justified, there must be some way to identify what the standard is. (If the standard cannot be identified, there's nothing for one to comply with.) But here the first signs of trouble emerge. Insofar as there's a problem about understanding rules, it is unclear how one would be in a better position to understand what a community agrees on as the standard for how to follow the rule. Evidently, if the agreement of the

[12] Here I agree wholeheartedly with Edward Minar who writes: 'To attribute a community agreement view to Wittgenstein underestimates the depth of his criticism of the whole idea [of a search for a "something" that fixes the interpretation of rules]' (Minar 1991, 206; square brackets by OK).

community is given a linguistic formulation, such as 'we meant you to follow the rule thus...' or 'the correct way to follow the rule is...', this cannot get us philosophically any further than the original rule, because the same problem that arose about rules arises about such formulations of the agreement of the community. Essentially, this appeal to community can do no more than give a second rule to explain how to follow a rule. Whilst such explanations and further rules are indeed the usual way to solve practical problems about how to follow a rule in everyday life, they cannot solve the philosophical problem of grounding because of the generality of this problem, as opposed to any particular practical problems.

Early on in his book Kripke expresses the problem with rules by saying 'what tells me that I should say "125" and not "5"?' (Kripke 1982, 21). I'm uncertain how exactly Kripke intends this, but it obviously cannot be understood literally without the original problem arising again. Whatever the community might tell Kripke, this is open to interpretation just like the original rule. But nothing ultimately depends here on how the community expresses its agreement about the rule, that is, whether this is done by means of words or actions. Whichever way the communal agreement is expressed, we are stuck with this question: why is the communal agreement itself not open to interpretation and misunderstandings like the rule? What extraordinary properties does it have that make it understood without any need for interpretation? As a person who has lived a significant part of his life abroad, my experience certainly is that understanding and following customs may be far from straightforward. This is relevant in that it helps to bring out the following. Insofar as acting according to a custom seems somehow straightforward and obvious to one, this is because one has internalized the customs in question. But to explain the possibility of rule-following in terms of internalized customs is question begging, provided Wittgenstein's characterization of rule-following as a custom (PI §199). We might just as well say that the possibility of rule-following requires one to internalize the rule. Clearly, this does nothing to address the problem, but merely restates it in different terms. (What one has internalized is something one knows, can do or understand.)

The problem just outlined is essentially a version of the original problem that arises between the pupil and teacher which Wittgenstein describes in *Investigations* §143ff. Here the teacher might naturally be taken to represent the agreement of the community. Indeed, imagining the whole community standing next to the pupil's desk and acting as a teacher is not going to somehow change the situation. Crucially, it is part of his example that when, after having passed the test of continuing the series +2 correctly up to 1000, the pupil is asked why he wrote down 1004, 1008, 1012, he looks at the teacher puzzled, declaring that he did do the same.

> Then we get the pupil to continue one series (say "+ 2") beyond 1000 – and he writes 1000, 1004, 1008, 1012.
>
> We say to him, "Look what you're doing!" – He doesn't understand. We say, "You should have added two: look how you began the series!" – He answers, "Yes, isn't it right? I thought that was how I had to do it." – Or suppose he pointed to the series and said, "But I did go on in the same way". – It would now be no use to say, "But can't you see…?" – and go over the old explanations and examples for him again. (PI §185)

The pupil, in other words, doesn't understand any better the teacher's pointing out that he did something different than he understands the rule itself. Here there seems to be no reason to think that he would do any better if the community was gathered around his desk instead of the teacher as its representative. As this brings out again, understood as some kind of authoritative voice, the agreement of community cannot do anything more than a second rule that might be used to fix the interpretation of the first rule. Just like the second rule is open to various interpretations, so is the agreement of the community, regardless of how we imagine it expressed. This is how we were led into the problem in the first place: any sign can be interpreted in multiple ways and any action is consistent with any rule suitably interpreted. Thus, unless it is simply stipulated that the agreement of community is not open to interpretation, the notion of communal agreement cannot help to explain rule-following. Instead of solving the problem, 'We keep on finding ourselves on the old level.' The problem about rules now simply repeats itself as a problem about communal agreement. This gives a reason to pause and rethink what role the communal agreement is supposed to play. For, clearly Wittgenstein does propose the notion of a practice of rule-following as a solution to the problem about rule-following *in some sense*. But we have yet to understand how this solution works.[13]

Let me quote a few more remarks in order to explain in a slightly different way the problem with the Kripkean solution in terms of communal agreement:

> One does not feel that one has always got to wait upon the nod (the prompt) of the rule. On the contrary, we are not on tenterhooks about what it will tell us next, but it always tells us the same, and we do what it tells us.
>
> One might say to the person one was training: "Look, I always do the same thing: I ..." (PI §223)

> The word "accord" [agreement; *Übereinstimmung*] and the word "rule" are related to one another; they are cousins. If I teach anyone the use of the one word, he learns the use of the other with it. (PI §224)

> The use of the word "rule" and the use of the word "same" are interwoven. (As are the use of "proposition" and the use of "true"). (PI §225)

Imagine someone saying, 'I know what rule-following is! It's always doing the same thing!' One could hardly object. To follow a rule is to do the same, for instance, to always add 2. Importantly, however, what counts as doing the same, depends on the rule. We have no notion of sameness for what counts as doing the same when following a certain rule that is independent of the rule. But if the notion of sameness is not given independently of the rule, then the notion of sameness cannot explain (in the sense of grounding) what it is to follow the rule. It already presupposes the rule, and cannot explain/ground it. Hence, insofar as the imagined someone wanted to explain what rule-following is by saying 'it's always

[13] For those attracted to the Kripkean account part of its appeal might be that the behaviour of the community seems to offer a non-semantical basis for explaining meaning, whereby rule-following and meaning are explained with reference to something grasped independently of our grasp of rules and meanings in the style of scientific explanations. As this section brings out, however, the notion of sameness relevant for rule-following is already a semantical notion. For what counts as doing the same as the community is not independent of the rule, but determined by it, and the same goes for using a word in the same meaning as a community does (see the next set of quotes from Wittgenstein (PI §§223–225)).

doing the same thing' she has not managed to do anything but to formulate the point in a different way: instead of saying, please follow this rule, one could just as well say, always do the same, namely, *this*. Further, as is easy to read out of the preceding remarks, the same point applies to the notions of agreement and accordance. To follow a rule is to act in agreement or accordance with other rule followers. That is indeed what it is to follow a rule, and trivially so. Such trivialities, however, cannot explain the possibility of rule-following. To merely state something in different terms cannot ground anything. What is the problem here more precisely?

In order to explain the possibility of something by saying 'the possibility of X depends on Y' (e.g. rule-following depends on communal agreement) it must be possible to identify the entity to be explained independently of what does the explaining. Otherwise the explanation is circular or tautologous, and merely states a triviality, as when saying that to follow a rule is to do the same, or that to follow a rule is to act in accordance with other rule-followers. These points are indeed true, but they cannot do any explanatory or justificatory work. To illustrate, they are crucially different from saying, for example, that in order for an object to move energy must be applied to it. Here we do have an independent grasp of what it is for an object to move, and the explanation explains how this is possible, thereby also excluding some other possibilities, such as an object moving by itself as if by magic. Similarly Wittgenstein's account excludes the possibility that intention fixes the meaning of a sign independently of the background of a practice (see PI §§197–199 and section "So What Is Rule-Following?"). But this account differs crucially from scientific explanations in terms of independently understood explanatory factors. As just outlined, the explanation of the possibility of rule-following or meaning by saying that their possibility depends on there being a communal agreement is merely illusory, landing us on the same old level as we started from. The question 'how is rule-following possible?' now becomes 'how is agreement with the community possible?' This then is the problem with trying to use communal agreement to explain what rule-following is. Communal agreement cannot ground or give a foundation for rule following, or justify any particular way of following a rule, because it is not independent of what is meant to be grounded, given a foundation for or justified. In the end the envisaged explanation is of this form: 'to follow a rule is to follow a rule = to act in agreement with other rule followers', or 'the possibility of rule-following rests on rule-following = on doing the same as other rule followers'.

This can be further clarified with the help of remarks which Wittgenstein makes on this kind of apparent explanations:

> "If humans were not in general agreed about the colours of things, if undetermined cases were not exceptional, then our concept of colour could not exist." No:—our concept would not exist. (MS 136, 121b/Z §351)
>
> "If we don't agree in/on certain things, we cannot argue"—Rather: without this agreement we simply don't call it arguing. (MS 118, 22r)
>
> If someone says "If our language had not this grammar, it could not express these facts"—it should be asked what "could" means here. (PI §497)

All three remarks (or at least the first two) make the same basic point. Insofar as something, for example agreement, is an essential feature of the practice, that feature cannot be appealed to as a condition of the possibility of the practice, because we have no grasp of the practice to be explained (for instance, the use of colour words) independently of the feature that was meant to explain it. And it is just in this way that talk about communal agreement brings to view something characteristic of rule-following. Rule-following is characterized by agreement; it is a matter of doing the same. But we cannot explain the possibility of rule-following or what fixes the rule or the meaning of the sign in terms of agreement of the community or accordance of the actions of rule-followers because this is just another face of rule-following. Here the contrast between accounts in terms of would and could marks a difference between clarifying the features of our concepts or relevant phenomena through description, for example by saying that rule-following is a practice or involves agreement, and offering empty pseudo-explanations of possibility.[14]

[14] See Kuusela (2008b) for how problems with transcendental arguments can be addressed from a Wittgensteinian point of view. Note that descriptions such as just mentioned constitute models to be used as objects of comparison. This enables one to address further muddles, for example, relating to the question how it is possible to change old practices or introduce new ones.

To further clarify the problem with the explanation of the possibility of rule-following in terms of agreement, compare this with another pseudo-explanation. The apparent explanation of rule-following is as little an explanation as it would be to say that the possibility of triangles having three angles depends on their having three intersecting sides. It is true that triangles have three angles, and this is intimately connected with their having three intersecting sides. But having three intersecting sides doesn't explain the possibility of having three angles because having three angles just is a matter of having three intersecting sides. We could just as well state the supposed explanation the other way around and this shows that it explains nothing, only restates what was to be explained.[15] What we have in our hands is a description, not an explanation (see section "So What Is Rule-Following?"). To pass off a description as a philosophical explanation of possibility results in an empty, merely apparent pseudo-explanation like the one about triangles—or rule-following in terms of communal agreement. To confuse such explanations with explanations in terms of independently understood explanatory factors is to confuse a respectable type of scientific explanation with merely apparent philosophical pseudo-explanation.

With regard to Wittgenstein's rejection of merely apparent pseudo-explanations of the preceding kind, there seems indeed to be an important difference between Wittgenstein and Kripke who writes, when outlining the solution attributed to Wittgenstein: '[…] if there was no general agreement in the community responses, the game of attributing concepts to individuals – as we have described it – *could not exist*' (Kripke 1982, 96; italics by OK). This formulation certainly looks very much like the kind of pseudo-explanations Wittgenstein rejects in the last three quotes. If this is how Kripke meant his explanation, it must be rejected on the same grounds as the other pseudo-explanations Wittgenstein mentions as examples.

[15] In PI §225 quoted above Wittgenstein mentions the notion of proposition and truth as interwoven in a relevant way. See PI §134ff. for rejection of the idea that the notion of truth can be used to explain what a proposition is.

So What Is Rule-Following?

Is Wittgenstein guilty of wasting our time with trivialities instead of giving a proper philosophical account of rule-following? What is the point of saying that rule-following is a practice or that rule-following involves agreement? Here we can start seeing that and how his discussion of rule-following is indeed consistent with his methodological remarks. Wittgenstein is offering us reminders with the purpose of showing a way out of a muddle, that is, the paradox that arises from the idea of rule-following is based on interpretation. His purpose is to render perspicuous what rule-following is instead of our entangling ourselves in unworkable explanations.

Recall Wittgenstein's argumentation strategy in the *Investigations* of going through all kinds of cases, such as a mental image or a formula coming to someone's mind, and showing that any image or formula, mental state, disposition or a sequence of actions can be interpreted in a variety of ways. The outcome of the argument was that none of these things can explain rule-following in the sense of determining the meaning of the rule or sign. The rejection of such an explanation, however, doesn't mean that things such as a mental image or a formula occurring to one or one starting to write a series down in a certain way could not count as instances of rule-following. Of course they can! The point is only that they cannot by themselves carry the weight of grounding rule-following or fixing the meaning of the rule, and explaining it in this sense. Rather, things such as a formula occurring to one require a practice of rule-following as their background or context. Against such a background or context a mental image or a formula coming to one's mind, and one continuing a series on that basis is a perfectly respectable instance of rule-following. A mental image or formula can indeed have a determinate meaning in such a case.

This point is illustrated by PI §197 on what it is to intend to play chess. I can express my intention to play chess by saying 'let's play chess' insofar as there's a practice of playing chess. Against this background, and the background of English, the sounds I make have a definite meaning and it is clear and determinate what game I'm proposing to play. Accordingly, my mind doesn't have to run through the rules to give the suggestion a definite sense; no peculiar processes are needed to explain the intention to play chess.[16] As Wittgenstein puts this point:

> An intention is embedded in its situation, in human customs and institutions. If the technique of the game of chess did not exist, I could not intend to play a game of chess. To the extent that I intend the construction of an English sentence in advance, that is made possible by the fact that I can speak English. (PI §337)[17]

Consequently, the outcome of Wittgenstein's rule-following argument can be stated as follows: rule-following is a variety of things, many different examples of which are mentioned in the course of Wittgenstein's rule-following discussion. It might be a matter of a mental image or formula occurring or just a feeling of 'yeah, I know how this goes' and doing what is asked. Or perhaps there's no mental occurrence or feeling but one just takes a breath and starts writing down the series whistling a Burt Bacharach song. Happily, rule-following therefore is what it seems to be. About this Kripke is right: rule-followers don't have to look over their shoulders and worry about the community and its acceptance (Kripke 1982, 79–80). On its own, however, if we isolate the moment of my writing down a number series from the history of the world and the practice of mathematics, what I write down is just scribbles. My intention cannot turn them into a mathematical series by itself. Nevertheless, against the background of the existence of mathematics, the fact that I spent many years learning it at school, and so on, what I write down and you assess for correctness can be understood as a series of Arabic numerals that satisfy the function you gave me.[18] I write down the series, you look at it and

[16] Note that on this account there's no problem with proposing to play a variant version of chess either. The ingredients for forming this intention, so to speak, are likewise available in the practices of language use and game-playing.

[17] Wittgenstein's 'could' in this remark is perfectly consistent with the earlier remarks about the difference between could and would. It is a different thing to say that one could not intend to play chess without the background of the practice, and that the practice itself could not exist without agreement.

[18] As Wittgenstein points out, mathematical education is relevant here, not as a causal history of how I came to be able to do calculations. Rather, the training I received at school indicates that there's a practice of doing these things, an established usage for mathematical signs and a custom (PI §198).

nod. That is all. Neither of us needs to check whether the community agrees. Rather, as our mathematical education indicates, we are part of this community, and we don't need to check with the community—unless a troublesome case arises and we need to consult others. But such cases raise no worries about the foundation of what we are doing. Even then the answer depends on mathematics. *The answer doesn't depend on any metaphysical fact about the agreement of the community.* It is simply characteristic of the practice of mathematics that there's an overwhelming agreement about how these things are done.

> Disputes do not break out (among mathematicians, say) over the question of whether or not a rule has been followed. People don't come to blows over it, for example. This belongs to the scaffolding from which our language operates (for example, yields descriptions). (PI §240)

> "So you are saying that human agreement decides what is true and what is false?" – What is true or false is what human beings say; and it is in their *language* that human beings agree. This is agreement not in opinions, but rather in form of life. (PI §241)

Yet again, the notion of form of life cannot be given a foundational role to explain the possibility of rule-following. It is characteristic of what mathematicians do, or of the form of life that doing mathematics is part of, that mathematicians agree on many things and never disagree on basic things such as how to add familiar kind of numbers or solve easy equations. (Without agreeing on such basic things they could not debate the truth of more complex matters. More generally, and I hope obviously: we cannot agree or disagree about the truth of what is said, unless we agree in the use of our words.) As explained, however, this agreement is not the foundation of mathematics, but a characteristic feature of it. There being such a practice and an agreement about how to employ mathematical symbols is important for rule-following, not as its foundation, but because this constitutes the background and context against which a variety of things, such as mental occurrences and writing down certain signs, can be understood as instances of rule-following.

Finally, it is worth noting that when this kind of clarification is given, no claim about what rule-following must be is made, and in this sense no philosophical thesis or theory is asserted. As I have argued, the explanation 'there could not be rule-following in the absence of communal agreement' is empty. Rather than offering us such an empty pseudo-explanation, the purpose of Wittgenstein's remarks on rule-following is to remind us about something we already know as rule-followers. What might look like a philosophical explanation concerning possibility of rule-following is a description of what rule-following involves.[19]

References

Kripke, S. 1982. *Wittgenstein on Rules and Private Language*. Oxford: Blackwell.

Kuusela, O. 2008. *The Struggle against Dogmatism*. Cambridge, MA: Harvard University Press.

———. 2008b. Transcendental Arguments and the Problem of Dogmatism. *International Journal of Philosophical Studies 16* (1): 57–75.

———. 2019a. On Wittgenstein's and Carnap's Conceptions of the Dissolution of Philosophical Problems, and against a Therapeutic Mix: How to Solve the Paradox of the *Tractatus*. *Philosophical Investigations 42* (3): 213–240.

———. 2019b. *Wittgenstein on Logic as the Method of Philosophy*. Oxford: Oxford University Press.

———. 2023a. Dissolution of Philosophical Problems and the Transformation of Thinking. In *Wittgensteinian Exercises: Aesthetic and Ethical Transformations*, ed. Lucilla Guidi. Leiden: Fink-Brill.

———. 2023b. Wittgenstein, Nietzsche, and Future Philosophers: The Notion of Truth in Philosophy. In *Wittgenstein and Nietzsche*, ed. Pascal Zambito and Shunichi Takagi. New York: Routledge.

Malcolm, N. 1986. *Nothing Is Hidden*. Oxford: Blackwell.

McDowell, J. 1984. Wittgenstein on Following a Rule. *Synthese 58*: 325–363.

[19] I would like to thank for questions and comments from those who attended Kripke's and Romina Padro's online seminar series on Kripke's *Wittgenstein on Rules and Private Language* in spring 2021 for which this chapter was originally written, as well as those who attended the *Wittgenstein and Practice* conference in Bergen in May 2022, where I presented the chapter for a second time. I'm grateful also to the editor of this volume for comments and suggestions.

———. 1992. Meaning and Intentionality in Wittgenstein's Later Philosophy. *Midwest Studies in Philosophy 17* (1): 40–52.
Minar, E. 1991. Wittgenstein and the "Contingency" of the Community. *Pacific Philosophical Quarterly 72*: 203–234.
———. 1994. Paradox and Privacy: On §§201–202 of Wittgenstein's *Philosophical Investigations*. *Philosophy and Phenomenological Research 54* (1): 43–75.
Wittgenstein, L. 1958. *Preliminary Studies for the "Philosophical Investigations" Generally Known as the Blue and Brown Books*. Oxford: Blackwell.
———. 1967. *Zettel*. Oxford: Blackwell.
———. 1989. *Wittgenstein's Lectures on the Foundations of Mathematics: Cambridge 1939*. Chicago: Chicago University Press.
———. 1993. *Philosophical Grammar*. Oxford: Blackwell.
———. 1998. *Remarks on the Foundations of Mathematics*. Oxford: Blackwell.
———. 2000. *Wittgenstein's Nachlass: The Bergen Electronic Edition*. Oxford: Oxford University Press.
———. 2009. *Philosophical Investigations*. Oxford: Blackwell-Wiley.
Wright, C. 2001. *Rails to Infinity*. Cambridge, MA: Harvard University Press.

Wittgenstein, Meaning and Action

Valérie Aucouturier

What We Say and What We Do

One of Wittgenstein's important moves in philosophy is to invite us to think of language as part of our human form of life rather than some independent domain of meaning separated from other aspects of our lives, such as what we do and why we do it. He thus invites us to rethink our 'natural history' (*PI*, §25) as including not only our biological condition as human beings but also our activities and regulated practices.

The concept of the 'language game' he uses to exemplify how language takes on meaning in relation to these activities is a philosophical tool which helps the philosopher clarify the conditions under which a word or a phrase can have meaning. But how can these activities by themselves instantiate meanings, if they do? If nothing grounds the rules of meaning outside our practices, nothing grounds the rules of meaning at all. And so there would be no rules of meaning at all. In this paper, I wish to explore

V. Aucouturier (✉)
UCLouvain Saint-Louis Bruxelles, Brussels, Belgium
e-mail: valerie.aucouturier@uclouvain.be

K. M. Cahill (ed.), *Wittgenstein on Practice*,
https://doi.org/10.1007/978-3-031-68656-6_5

how acknowledging that the meaning of expressions is dependent on human practices does not entail that our possession of concepts is (Anscombe 1976, p. 121).

To begin with, here is one way, for Wittgenstein, of understanding how language games can take root in what we do:

> The origin & the primitive form of the language game is a reaction; only from this can the more complicated forms grow. Language – I want to say – is a refinement, 'In the beginning was the deed'. (Wittgenstein, *CV*, 36)

In this passage Wittgenstein shares Faust's disbelief in the magical power of the 'word' and its ability to produce meaning by itself. Words take meaning in action, uses and language games. Meaning (language) is inseparable from what humans do. It is therefore necessary to abandon the old dichotomy between language and the world, and to understand language in relation to action.

On the one hand, it makes no sense to try to distinguish, in what we do (our actions), what would be a result of some 'natural' causality from regulated actions. There are indeed clearly regulated actions (institutional acts, a number of speech acts are most typical examples), but the porosity of actions and language and conventions makes it impossible to distinguish sharply between so called natural gestures (e.g. the expression of 'pure' emotion) and conventional actions. On the other hand, language is built on practices and is anchored in practices of which it is part, whether these are strictly conventional (such as promising or making a commitment, or the rules of politeness) or not (such as getting angry or expressing pain or shame)—if such a distinction makes sense.

This embeddedness of meanings in practice and action calls for a certain naturalistic reading of Wittgenstein, found, for example, in Stanley Cavell (2015) or Sandra Laugier (2008). That sort of naturalism should not be conflated with some 'bald naturalism', placing all phenomena in the world under the reign of the laws of nature. Rather, it suggests that language games can grow on some more 'primitive form', which is, among other things, that of 'reaction'. Thus, a cry of pain may become 'I am in pain'; a gasp, 'I am afraid', etc. Emotions are undoubtedly one privileged place where this continuity between natural reactions and conventional expressions is exemplified.

However, Wittgenstein's naturalism, taken in this sense, is not restricted to the continuity from primitive reactions to language games. It also implies a continuity between more structured actions and language games. Language games are indeed not just expressing in words or substituting words for our natural or primitive reactions, they are also part of the more general domain of what we do, our actions.

> Giving orders, asking questions, telling stories, having a chat, are as much a part of our natural history as walking, eating, drinking, playing. (*PI*, §25)

Nevertheless, as we shall see, this naturalism is no ground for a theory of meaning. It is the mere acknowledgment of the entanglement of language into what we do, which constitutes our forms or form of life.[1] This entanglement is part of the notion of 'language game', even though this notion emphasizes the regulated character of the uses of language—justifying the appeal to the idea of a 'philosophical grammar' or logic.

On the Game Metaphor: Rules and Actions

The concept of language game bridges language and practice through the process of learning, which first takes the form of a training (*Abrichtung*). To describe the meaning of a word, the first appropriate question to ask is how we could have learnt it: under which practical conditions can an individual acquire the use of a word, a sentence or an expression? This diversion from the general question, 'What is the meaning of a word?', to the question of the conditions under which one can learn to use a piece of language is precisely what allows us to re-examine the matter in its practical dimension. It also makes it possible to express the conditions of possibility of the learning of a word, which constitute the characteristic features of its 'grammar'.

Thus, in a certain sense, we can say grammar expresses or exposes the conditions of possibility of meaning. It exhibits the conditions enabling

[1] The form(s) of life can be said in the plural or in the singular depending on whether it refers to human 'natural history' or to more local aspects of our practices.

a word, an utterance, a speech act, to make sense on a given occasion. But these conditions do not come in the form of a list of (necessary and sufficient) conditions that must be met to make language operational or meaningful. Insofar as meaning is necessarily linked to an occasion of speech, the uncovering of its conditions requires the description of uses, i.e. actions or practices, which bring words, statements or speech acts into play. It is in the light of the examination of language games that one can test the limits of meaning, i.e. the places where language loses its footing, where use has been lost. Cases where one would be unable to describe a practice or a language game in which the words in question take on meaning, where one can no longer attach a usage to a practice, namely a pseudo-use; a use that is not (yet) a use.

However, the notion of 'language game' cannot either by itself characterize everything we do in and with language. For, if it emphasizes that speaking is a regulated practice, it does not exactly account for the 'naturalness' of language, i.e. its continuity with behaviors or attitudes, such as a cry of pain or some expression of feelings. Therefore, the descriptive power of grammar must be put into perspective.

Philosophical grammar gives an account of the philosopher's activity in untangling conceptual quandaries. It does not however provide an exhaustive description of the functioning of language. Nevertheless, it raises the issue of the conditions of possibility of agreement on rules and in language. For the functioning of a rule or a game seems to rest primarily on the possibility of an agreement on what it is to follow a rule. Otherwise, how could we guarantee we are actually playing the same game and not merely appearing to be doing so?

Wittgenstein famously made such an objection to himself at paragraph 201 of *Philosophical Investigations*:

> This was our paradox: no course of action could be determined by a rule, because every course of action can be brought into accord with the rule. The answer was: if every course of action can be brought into accord with the rule, then it can also be brought into conflict with it. And so there would be neither accord nor conflict here. (PI §201)

The rule-following issue takes on the appearance of a vicious circle. The rule—the one, among others, involved in our language games, which guarantees that we are saying and meaning something—cannot logically precede an action that is supposed to constitute or consist in following the rule. But then, if there is no necessity of the rule, there appears to be no absolute obligation to follow it. And if there is no necessity to follow the rule, there is no guarantee that we are following it, even when we seem to be doing so. If the rule must both constitute the agreement and make it possible, it loses its 'logical rigidity'. This skeptical objection was knowingly taken a little too seriously by Saul Kripke (1982), who concluded that following a rule must therefore be arbitrary.

No Ground for the Rule

Kripke's Wittgenstein, known as 'Kripkenstein', understood the rule-following paradox as an inextricably skeptical argument: when I speak or operate according to a regulated activity (like calculating in mathematics), there's no guarantee that I am following the rule correctly, even if everything I do corresponds exactly to what I would do if I were following it. 'After all, whatever I do can, on some interpretation, be made compatible with the rule' (*PI*, §198), says the objector to Wittgenstein.

Indeed, if, as pointed out above, the only thing that guarantees following the rule is practice, how can we guarantee that a practice actually consists in and does not merely resembles following a rule? Rules do not by themselves account for their application; they are not 'rails' that we follow blindly. But does that mean we are merely 'interpreting' them, for example, by looking at what others are doing?

Kripke's solution to this paradox consists in calling on the notions of community and life form to guarantee some sort of agreement on the rules. The rule paradox would only arise for the one who follows the rule 'in private', outside the linguistic community—i.e., who ultimately follows no rule at all, since following a rule in private amounts to modifying rules as one gets along.

In fact, Wittgenstein's point is much more radical. Kripkenstein's mistake is to maintain a foundationalist reading of the rule-following paradox. It forgets the radicality of the observation (for it is an observation, not a philosophical thesis) of the primacy of action. In other words, to ask the question of the origin of agreement is to ask the wrong question. Even more so, as Laugier points out, it 'even constitutes the skeptical question' (Laugier 2008, 51).

The immanence of practice and custom, our immersion in them, prevents us from seeing the problem of agreeing with or following the rule as that of a simple adherence of the individual or even a community of individuals to some rule, suggesting the possibility of extracting ourselves from it to think it outside of practice. I do not face the rule, however, I practice it. Our understanding of the rule is not external to the rule. Hence the importance for Wittgenstein of understanding the learning process as a form of training. To understand the rule is not first to intellectualize it, but to be able to incorporate it into one's actions.

This rejection of the communitarian solution to the problem of skepticism (or relativism) consists in a rejection of the notion of a background—of practice as background (found, e.g., in Searle 1983). Indeed, thinking the form of life as a 'condition of possibility' or 'background' gives the false impression that there is first some 'natural' form of life against which agreement is possible, and on to which convention is grafted. It leads to the transcendental illusion that we could extract ourselves from our form of life in order to think it through, that we could understand the form of life, among other things, without the very medium of this understanding, namely language. However, agreement is the form of *life*, as Laugier (2008) emphasizes with the word 'life', i.e. a condition that is not itself in question. The rule is not grounded on the practice of language games, it is that very practice. To exhibit the rule is to *describe* a practice, no more, no less.

Conditions of Fit

However, these discussions about rules—which exhibit the entanglement of language and action—leave the question of the conditions of agreement open. Indeed, we do not agree on rules but *in* language and practice. If the question of the *foundation* of agreement is unintelligible, we can nonetheless ask the question of the *conditions* of agreement.[2]

These considerations raise the issue of the boundary between language and the world. If we cannot extricate ourselves from our practices, if we certainly cannot say anything outside language, does this imply that language is incapable of speaking of anything outside itself? The observation that we cannot disentangle what we do and what we say may sound as the rejection of any kind of distinction between the 'conceptual' and the 'non-conceptual', i.e. between what belongs to language and what does not.

John McDowell (1996), for instance, seems to think that questioning the dichotomy between language and the world also calls into question the boundary between the conceptual and the non-conceptual. According to him, our perception of the world is tinged with concepts. If there is nothing we can reach independently of a conceptual system that shapes our world(view), if there is no object of thought outside language, then it seems logically and empirically impossible to distinguish an independent reality grasped by our concepts. Correlatively, if every practice is interwoven with concepts, if language is inextricably intertwined with practices, what sense could there be in thinking about language separately? A misunderstanding of paragraph 241 of *Philosophical Investigations* could induce such a reading:

> 'So you are saying that human agreement decides what is true and what is false?' What is true or false is what human beings say; and it is in their language that human beings agree. This is agreement not in opinions, but rather in form of life. (*PI*, §241)

At first glance, it might seem Wittgenstein is countering the objection to the arbitrariness of judgment (and, by extension, of rule), which states that if humans do decide what is true and what is false, then there can be no objectivity of judgment. To this end, he seems to appeal to the concept of a 'form of life' as the basis of judgment. In one sense, a form of life

---

[2] Which leads some readers of Wittgenstein to say that, even if he rejects any transcendental posture, he does not totally abolish every transcendental questioning of the logical—not empirical—conditions of possibility of meaning.

corresponds to the minimum shared experience that makes agreement and communication between humans possible. But to try to ground agreement on something like a form of life would amount to denying the radical immanentism of practice and calling into question the rejection of the very relevance of the issue of foundations. It is indeed *in* language that the source of agreement is to be sought. But that is not to say that language is incapable of reaching outside language. Quite the opposite. To understand this, let us continue reading paragraph 242:

> It is not only agreement in definitions, but also (odd as it may sound) agreement in judgements that is required for communication by means of language. This seems to abolish logic, but does not do so. It is one thing to describe methods of measurement, and another to obtain and state results of measurement. But what we call 'measuring' is in part determined by a certain constancy in results of measurement. (*PI*, §242)

If our concepts can fit a situation, be perfectly in tune with it, or even be entirely appropriate, it is not by virtue of some magical relationship they share with the situation. It's precisely because they themselves are the measure of that fitness. It would make no sense to measure the yardstick to see if it is the right measure, because the yardstick provides the criterion for the right measure. Similarly, our concepts do not need to be compared with anything else to guarantee that they say what they mean, because they are the very measure of what they mean.

However—and this is where we must not lose sight of the practical necessity that gives rise to concepts—there is indeed something that a method measures, and this measurement has ends or reasons for being. It appears to be in this interstice between the method of measurement and its uses (between concepts and their uses) that 'agreement in judgment' comes into play. If there are indeed rules for the uses of language (even when these uses are irreducible to the enunciation of rules), we need to be able to agree on these uses. Not explicit agreement, but rather, as Wittgenstein puts it, a certain 'constancy' or stability that makes such agreement possible. If reality itself does not dictate methods of measurement, these methods must nevertheless be commensurate with reality. How could they not be? This is only possible insofar as these methods are directly built into practices (and sometimes even practical necessities). So, there is indeed a certain autonomy of logic, an autonomy of meaning, which, like a method of measurement, shapes our perspectives—understood as some sort of objective point of view on reality, detached from the subject.

On the other hand, if language is fundamentally human, if its capacity to signify depends on human existence and practices, this does not mean that it should be understood as conditioned or relative to some generic human 'point of view'. For it is precisely this idea of a view from the outside, or an outside of signification that would ground it, that Wittgenstein challenges.

The autonomy of the conceptual (to borrow a phrase from Charles Travis 2010, 277) is not to be understood as a possibility of extracting ourselves from language in order to understand its logic—as if we could look at the structure of our field of vision. There is no meta-perspective from which to understand the relationships between the various perspectives, but neither are we locked into our perspectives or our points of view. Grammar is not a question of perspective, even if language games do sometimes draw up views of the world, and even if we cannot think outside language.

Wittgenstein's Partial Idealism

This last remark, on the anthropological dimension of language games as characterizing forms of life, or perspectives on the world, suggests a way to clear up a misunderstanding, before concluding. The anthropological dimension of language raises the question of linguistic idealism as understood by Elizabeth Anscombe (1976).

In her paper on 'The Question of Linguistic Idealism' (Anscombe 1976), she distinguishes between two ways of understanding linguistic idealism: the first is logical, the second is anthropological. The logical question asks whether we can distinguish the logical form of a phrase from the reality the phrase speaks of. In other words, to what extent do our linguistic categories shape our environment? The anthropological question is whether it is possible to compare systems of thought (or knowledge) and say that one system could account for the wrongness of another, or whether we are instead condemned to think within incommensurable linguistic categories, language games and systems of rationality, and are irremediably caught up in the visions of reality that stem from them. To assess the role of practice in Wittgenstein's understanding of meaning, we must address the logical question of the extent to which language shapes reality and makes certain things exist in it and the related anthropological question of the incommensurability of worldviews.

Anscombe's (1976) article opens with a quote from Wittgenstein's *Philosophical Investigations* about the contingent nature of our concepts:

> If anyone believes that certain concepts are absolutely the correct ones, and that having different ones would mean not realizing something that we realize, then let him imagine certain very general facts of nature to be different from what we are used to, and the formation of concepts different from the usual ones will become intelligible to him. (*PI*, II, xii)

The issue of linguistic idealism lies in the articulation of three aspects to be found in this passage: our concepts (the logical categories we use to speak and think), our conceptions (what we 'realize', the way we conceive of what we speak and think about) and 'very general facts of nature'. But to what extent do very general facts of nature constrain our linguistic categories and thus inform our conceptions? To what extent do concepts influence our conceptions and understanding of the very general facts of nature? In this respect, Wittgenstein's remark is clear: very general facts of nature ('regularities') constrain our concepts (and categories) and consequently our conceptions. However, our concepts are not 'absolutely the correct ones'. They are only possible, not necessary, ways of categorizing 'very general facts of nature' and informing our conceptions of them. As such, our concepts are partly contingent. They could have been otherwise.

But if, as Wittgenstein suggests, our concepts could have been different from what they are, and if we are to conceive that they could be, how are we to think of the relation of our concepts to those 'very general facts of nature'? Moreover, to what extent are we to think of our concepts as conditioning, so to speak, our perspective on the world (our conceptions) and thus constituting, in a sense, the 'essence' (Wittgenstein *PI*, §371) of the world we live in? To what extent do our concepts shape our conceptions of the very general facts of nature? To what extent do very general facts of nature lead us to adopt or create certain concepts?

The accusation of linguistic idealism is thus an expression of suspicion toward the contingency of concepts and toward the criticism of the claim that our concepts would be ideal or isomorphic reflections of facts of nature. We saw that such contingency cannot be purely arbitrary. Facts of nature do constrain the adoption of certain concepts. It is therefore important to maintain the Fregean (Frege 1892) distinction between concepts, their meaning, and what they are about.

Still, Wittgenstein criticizes the idea that concepts or linguistic categories constitute an 'absolutely correct' image of reality, the right image of reality or of what there is to be thought, or at least that these concepts tend to be so. He thereby criticizes a certain empiricist approach to meaning (found, e.g., in John Locke) according to which reality or rather the experience of reality could by itself prescribe linguistic categories. Reality can indeed prompt us to forge new concepts; and forging new concepts is also to imagine a different 'world'.

Starting from this difficulty, Anscombe distinguishes two types of idealism:

> So now it looks as if *either* the grammar corresponded to something of the object, its real essence, which it has whether there is language about it or not, *or* the 'object' were itself dependent on language. The first is like the suggestion made by Plato in the *Cratylus*; the second, if it applies through and through, I call 'linguistic idealism'. (Anscombe 1976, p. 113)

The Platonic idealism from *Cratylus*—which really is just a way of essentializing grammar by making its logical form a transcendent necessity independent of thinkers—ought to be rejected. According to this picture, grammar would correspond to the real essence of things, independent of language itself. Linguistic idealism, on the other hand, would be the idea that (the existence of) what language speaks of 'depends on language'.

Now, if we can possibly speak of idealism with regard to Wittgenstein's grammar, it is in the sense that the rules of grammar express essences but do not themselves exist independently of human practices. This is what Anscombe calls *partial idealism*. The one thing we do not want in linguistic idealism is the idea that 'what we realize' would be a pure and mere projection of thoughts onto reality that would not exist independently of thoughts. In other words, we do not want to imply that nothing would have been red without human language, that is, without the linguistic practice in which the existence of the word 'red' manifests that its users possess the concept of red. The difficulty is that, on the one hand, we do not want to say that mere experience could dictate the use of a word (in the manner of Augustine's ostensive definition or the Lockean conception of ideas—see *PI*, §1). On the other hand, we do not want to imply that essences *expressed* by grammar are simply *created* by grammar.

By saying that 'essence is expressed in grammar', Wittgenstein (*PI*, §371) points out that the study of grammar, of the uses of words in language games, exhibits conventions (Anscombe 1976, p. 113) by means of which we grasp (talk of) reality. As Anscombe puts it,

> [When we say that] 'even if there had never been any human language so that there *was* "no concept of pain" at all – still if there were animals there would have been pain' [...] '[i]t is pain as we mean 'pain' that we are saying would have existed anyway'. And similarly, if there never had been humans around talking about horses, that is not the slightest reason to say there wouldn't have been horses. (Anscombe 1976, p. 114)

Thus, the essences expressed by grammar, and which constitute our 'conceptions', are not created by grammar, even if grammar itself is dependent on human practice. Not having the same grammar does not necessarily mean lacking something. Although it is possible to miss differences by lacking concepts to grasp them, this possibility is contingent. It depends precisely on the possibility of acquiring a concept.

Of pain or of horses, we can say they would have existed independently of the existence of human language, but this does not impose any necessity on the way in which the concepts of pain and horses categorize what they speak of. There are and could be indefinitely many ways of categorizing (e.g. colors, animals):

> The essence is expressed by grammar. But we can conceive of different concepts, i.e. a language that does not have the same grammar. The people who would use it would not be using a language whose grammar would express the same essences. However, we cannot conclude that they would fail to conceive something that we conceive. (Anscombe 1976, p. 133)

Taking the third way between idealism and empiricist realism (Wittgenstein, *RFM*, VI, §23) entails that uses of language cannot depend solely on representations:

> 'No image or representation could determine future or past application of the word'. But 'the essence [expressed by grammar] is not what I mean or am speaking of: it is rather that through which I understand or think of (mean), etc. That is to say, is that because of which my use of the word is a case of meaning a kind of animal'. (Anscombe 1976, p. 115)

This idea was already partly present in Frege's (1892) conception of meaning as a path to reference, which is not constituted by a psychological act of meaning, but rather through shared meanings (with criteria of 'shareability').

One way of understanding the logical requirement of public and shareable dimensions of meaning (which refers back to Wittgenstein's famous critique of the possibility of a private language) is to emphasize (at least for a number of cases) the importance of the method of 'counting' what is being talked about. Sometimes, to grasp a concept is to understand how to count what falls under that concept. And an ostensive definition cannot on its own determine what is being counted, nor determine a method of counting (which is indeed fixed by grammar).

In the cases considered so far (horses and pain), *what is counted* is not a creation of the mind. What possibly is a creation of the mind is the grammar of the expression, for example *the method of counting* (which characterizes the concept and which could have been different).[3] The idea or essence is therefore neither a pure creation of the mind, nor an independent object of thought, it is that by which the grammar of a word

[3] The method of counting appears thus as a possible (but not the only) way of expressing something that we realize.

makes it possible to speak of horses and to think of a horse by means of language. It is precisely this which is or can be common to different words in different languages. And this takes us back to the issue of rule-following and agreement.

Horses and pains are obviously not the only kinds of things we can reach through concepts. Coins and institutions (such as rules) are (or so it seems) more refined examples of realities, the existence of which is itself *dependent* on humans. Language itself is another specific example. This is the reason why, based on these considerations, Anscombe proposes the following test to identify a form of linguistic idealism she calls 'partial idealism':

> We shall ask the question: Does this existence, or this truth, depend upon human linguistic practice? That the *meaning of expressions* is so dependent is evident; that human possession of concepts is so dependent is not quite so evident. (Anscombe 1976, p. 116)

The meaning of words depends on linguistic practices, but the possession of concepts does not necessarily. To possess a concept is, among other things, to be able to make competent use of it. However, the possession of concepts implies activities other than simply speaking; activities which take place in circumstances that do not exclusively involve concepts.

> Humans would not, I suppose, have had the concept of the natural numbers if they had not had the practice of counting objects. (What do I mean by this? I suppose something like the following: if there isn't a practice of counting objects, it is difficult to see how anything could be identified as a series of natural numbers). But this fact does not dispose us to think that there could have been no such event as that a wolf killed three deer in seven days, before there were humans with their linguistic practices. (Anscombe 1976, p. 117)

The right distinction—leading to the idea of partial idealism—appears to be that some parts of the grammar (its rules, one might say) are human creations, though they are linked to practices and events whose existence does not exclusively depend on grammar. Birds build nests, but do they need a concept of bird-nesting to do so? Lions hunt, but do they have a grammar of hunting? Newborns eat, sleep and cry, but do they need to possess the concepts of eating, sleeping and crying to do so? There are lots of things we do without requiring any grammar for it. Grammar merely exhibits or describes the rules according to which we organize our (linguistic) practices. '[Concepts] are the expression of our interest' (Wittgenstein *PI*, §570)

> Humans have invented such a thing as 2 x 2, though they didn't invent 2. And though they didn't invent halves, they invented 1/2. They invented doing that with the ideas of *one* and *two* (there was no similar representation of a half in the Roman numeral system) and they invented treating it as a representative of half of anything. They invented treating half as a number. You might say they invented half's *being* a number. – Probably this is the right direction to look in.
>
> And now this leads us on to the possibility of what we might call a partial idealism. We have agreed (with small reservations) that the existence of human concepts can be somewhat generally equated with the existence of a great variety of human linguistic practices. But that, we have remarked, by no means implies any dependence on human thought and language, on the part of the things that fall under these concepts. (Anscombe 1976, pp. 117–118)

Not all the things that fall under the concepts we use, not all the things we can describe or account for in human language, are dependent on human thought and language. But some are. This partial idealism, which concerns what depends for its existence on human practices, is illustrated by the case of rules (including the rules of grammar), rights and promises.

Anscombe starts her analysis of rules, rights and promises by sharing David Hume's observation that promises, and more particularly the commitments they produce, are 'naturally unintelligible' (Hume 1975, III, ii, 5). *What* the rule allows or forbids cannot be considered as a mere description of behavior or of some natural necessity or constraint. The necessity of the rule (in this case the commitment it generates) derives from the existence of the rule. And the existence of the rule depends on our practices. However, there is a difference between the promise (as a linguistic or verbal sign) and the rule: the rule is the *meaning* of the promise, the grammar of the sign. It is not the statement of the rule that imposes any obligation; it is its meaning. It is in this sense that it is not a natural necessity.

Thus, while horses, giraffes, colors, shapes are not the product of human linguistic practices, the rules of logic or grammar, i.e. the necessities regarding the essence of things, are the product of human practices (Anscombe 1976, p. 121).

Yet grammar is not arbitrary. What determines the meaning of the rule is what we do. And this is also true of the rules of reasoning. Kripke was right in noticing that the rigor of logic does not come from outside. It is the result of a linguistic creation. The system of logic is a human construction. But it is objective in the sense that it does not depend on one individual's actions but on human practices (including linguistic practices).

The Anthropological Question

Now it is the contingency of practices and of what regulates them (and their sensitivity to circumstances) that raises the difficult question of the commensurability of systems of thought and the universality of logic. It confronts us with the absence of external rational bases for judging other practices and beliefs.

> As readers of the 'Notes on Frazer' will know, Wittgenstein rejected the idea that 'our science' shows that magical practices and beliefs are errors. Above all, he thought it stupid to take magic for mistaken science. Science can correct only scientific error, can detect error only in its own domain; in thoughts belonging to its own system of proceeding. (Anscombe 1976, p. 125)

Wittgenstein saw, however, that the absence of a meta-truth or meta-judgement pointed in this quote should not be conceived as a lack or as something to be sought outside or below human practices.

This difficulty reaches debates about Wittgenstein's possible epistemic relativism, understood as the impossibility of conceiving of a truth that is not simply local, relative to situated ways of conceiving the world, to systems of thought or language games that are likely to be incommensurable with each other. These debates concern the status of a number of shared minimal beliefs—or rather 'certainties' (*Gewissheiten*), to echo Wittgenstein's last text—that make the edifice of knowledge possible. These shared certainties, while themselves situated, would make our situated perspectives on the world commensurable.

Without entering a debate on the status of these certainties that stabilize our *Weltanschauungen*, our conceptions, we should simply avoid the confusion associated with this discussion and emphasize that we must distinguish the question of the term of justification (e.g. of scientific statements) from the philosophical question of foundation. In other words, we need to distinguish between cases where the demand for justification stops because it is no longer relevant (it does not apply and receives neither right nor wrong answers) and those where we seek to get out of a system in order to judge its rationality, when there are no criteria of rationality (enabling comparisons, agreements and disagreements) outside the system. In the context of justification, some statements are accepted as certainties, because they form the basis of a worldview (*Weltanschauung*). Now, it's one thing for a worldview to be conditioned by certain fundamental beliefs, which could eventually come to be called into question; but to export the requirement of rationality outside the logical system within which it takes on meaning is an aberration or a logical impossibility.

The vanity of the quest for foundations does not, however, point to a limitation of our knowledge (as Kant might have thought, for example), but to the emptiness of the quest itself. The quest for foundations is vain because it is not a quest. It is not a quest for something unattainable (due to the limits of our understanding), it is not a quest at all, since neither the modalities nor the object of the quest can be defined. This is a logical limitation not a limitation of the scope of human understanding. It is in this sense that logic traces the limits of the thinkable: not in the sense that thought would be limited (to a language system, e.g., which merely instantiates an empirically contingent possibility: for example perhaps certain languages prevent us from seeing or thinking certain things, and vice versa), not in the sense that there would be an outside, a point of comparison from which to think this limit. But logic traces the limits of the thinkable in the sense that we cannot extract ourselves from it to consider it from 'sideways on' (McDowell 1996).

The critique of the relevance of the notion of a viewpoint for thinking the commensurability of language games is therefore distinct from the critique of a relativism of forms of life *per se*. Agreement between humans is not constructed *ex nihilo*, but always in language (Wittgenstein, *PI*, §241). And even when systems of thought (e.g. beliefs) seem to be at a standstill, it is not necessary that irreconcilable systems of thought are irreconcilable throughout, but only in certain respects. We may teach a child to believe in God or not to believe in God (Wittgenstein, *OC*, §107), and this may have important consequences on her conceptions. A believer might well be unable to convert an atheist and conversely. But they can still speak, agree and disagree on some matters. Such a distinction between a believer and an atheist relates to what Travis calls 'the parochial':

> I will refer to as 'parochial' any form, or shape, of some being's thought which is not required simply by the demand of being a thinker as such, so that there is room for there to be thinkers whose thought lacked that feature—so to any form of thought for which there might be others. So, too, any feature of those generalities under which we are equipped to bring things (present things to ourselves) which might be absent from the stock of generalities under which some other sort of thinker is equipped to bring things will be parochial in the present sense. (Travis 2010, p. 1)

The parochial is that which, while having the generality of the conceptual, can be specific to a particular thinker or group of thinkers and need not be a trait common to all thinkers in general. Travis's understanding of the concept of the parochial involves the notion of 'occasion sensitivity'. The parochial is the dimension of language where the generality of the conceptual meets the absolute singularity of the context, where the general comes to grasp the particular. What Travis calls 'sensitivity to occasion' thus lies in our ability to make words say what we want them to say, and to grasp what our interlocutors say to us in context. And this sensitivity to occasion comes precisely from what Travis calls 'parochialism', a kind of understanding produced in and by language in a very local way, which enables us to relate what we say to particular usages. It is precisely this parochialism that is lacking in strictly empiricist conceptions of meaning:

> Cast in present terms, the aim of empiricism is to deny the parochial any role in making thoughts available to us, or at least to limit that role to the minimum, so that, at least very nearly, it is the world itself, and nothing else, which makes generalities available to us in terms of which to think. (Travis 2010, p. 4)

The role of the parochial, which is constituted in practice, is precisely to make it possible to contextualize the meaning of our language games. But once again, the common point of view in which this parochialism might consist is rooted in practice, in the contextual uses of language. Therefore, the parochial does not pertain to a private character of meaning. The parochial is no solipsism.

We make language say what we want it to say, not in the sense that it allows us to externalize or express internal, subjective meanings. We make language say what we want to say, in that we allow it to reach the particular in the articulation between a context and a sensitivity to the way our words can reach certain dimensions of the context. This is why the singularity of a point of view or perspective is to be distinguished from the singularity of context. My point of view may be unique in the sense that one would have to be me to take it, but the problem of meaning is not a matter of point of view. Meaning is a matter of commonality, to which the parochial belongs even in its most idiosyncratic aspects.

Perhaps, to some extent, the 'form of life' points to the existence of a common ground on the basis of which minimal forms of agreement are possible. The mistake, however, would be to consider it as a *foundation* of agreement.

References

Anscombe, Elizabeth. 1976. The Question of Linguistic Idealism. In *From Parmenide to Wittgenstein*. Oxford: Basil Blackwell, 1981.

Cavell, Stanley. 2015. *Must We Mean What We Say*. 2nd ed. Cambridge: Cambridge University Press.

Frege, Gottlob. 1892. Über Sinn und Bedeutung. In *Funktion, Begriff, Bedeutung. Fünf logische Studien*, 38–63. Göttingen: Vandenhoeck & Ruprecht, 1962.

Hume, David. 1975. *A Treatise of Human Nature*. Ed. L.A. Selby-Bigge, Rev. P.H. Nidditch, 2nd ed. Oxford: Clarendon Press.

Kripke, Saul. 1982. *Wittgenstein on Rules and Private Language*. Cambridge, MA: Harvard University Press.

Laugier, Sandra. 2008. Règles, formes de vie et relativisme chez Wittgenstein. *Noesis* 14: 41–80.

McDowell, John. 1996. *Mind and World*. Cambridge, MA: Harvard University Press.

Searle, John R. 1983. *Intentionality*. Cambridge, MA: Cambridge University Press.

Travis, Charles. 2010. *Objectivity and the Parochial*. Oxford: Oxford University Press.

What Is a Context?

Martin Gustafsson

Introduction

One point made by saying that language is a "practice" or "custom" is that *repetition* is essential to our use of words. Learning to speak means learning to do and take oneself to be doing *the same thing* again and again—to perceive and follow patterns of continuity. We make ourselves intelligible in speech by relying on established habits of usage.

However, these patterns of continuity are not nailed down once and for all by strict, inflexible rules, but allow for spontaneous variability and development. I take one of the central themes in Wittgenstein's later philosophy to be the interplay between repetition and variability in language. And I think many quarrels among philosophers influenced by Wittgenstein involve fundamental disagreements about how to conceive this interplay.

One aspect of the variability is the *context-sensitivity* of what our words are used to express. Wittgenstein is often said to have held a view

M. Gustafsson (✉)
Åbo Akademi University, Turku, Finland
e-mail: Martin.Gustafsson@abo.fi

K. M. Cahill (ed.), *Wittgenstein on Practice*,
https://doi.org/10.1007/978-3-031-68656-6_6

according to which context-sensitivity is pervasive—a sort of view which is nowadays called "radical contextualism".[1] According to radical contextualism, context-sensitivity is not limited to a some supposedly well-behaved subset of words such as indexicals and demonstratives, but characterizes our use of *all* linguistic expressions. The upshot of such radicalism is that context-sensitivity cannot be eliminated by adding elements to our sentences in order to adjust for indexicality, ambiguity, ellipsis, and so on. Such adding can never, even in principle, reach a point at which what such an extended sentence can be used to say is fixed independently of the particularities of the communicative situation.

A common worry about such radical contextualism, both as a reading of Wittgenstein and as a view in its own right, is that it conceives linguistic practice as much too disintegrated to be a linguistic practice at all. No one denies that what words are used to say is to some extent sensitive to variations in circumstances of use, but, the objection goes, isn't semantic stability across contexts at least as striking and significant? Indeed, isn't a fundamental feature of our meaningful use of words that what we can say in using words on a certain occasion depends on what those words mean regardless of how that occasion is different from other occasions of use? As John McDowell puts it, we should not

> abandon the thought that the primary form of the ability to mean something by verbal behavior is the ability to mean what one's words mean, independently of the particularity of one's communicative situation – that is, what they can mean in the language, in the ordinary sense, that one is learning to speak. (McDowell 2002, 187)

Or, as Jason Bridges has put this idea in a manuscript:

> [W]e natural-language speakers suppose that the words of our language are apt, all by themselves, for providing the content of a claim, and that in uttering them assertively, we aim to help ourselves to that expressive power – to let our words do the talking, as it were. (Bridges MS, 15)

---

[1] Charles Travis is perhaps the most influential proponent of such an interpretation. See Travis 1989.

Now, what, exactly, does this notion of continuity or stability across contexts involve? Importantly, both McDowell and Bridges find the commitment to such continuity or stability to be *part* of our ordinary linguistic practice, and not just a theoretical assumption imposed on language by philosophers or linguists. So, the question is, how does the commitment to such stability or continuity show up *in* the practice itself? I want to focus here on one reason to find such a commitment at work in ordinary linguistic practice, namely what Bridges has called "the homophonic report principle". He formulates this principle as follows:

> adjusting for indexicality, tense, and the like, we can correctly report what someone said on an earlier occasion with a "that"-clause that simply repeats the sentence originally uttered. (Bridges MS, 15, n. 17)

Here is an example which illustrates this point. A murder has taken place, and the police arrive at the scene within seconds. There is one witness who seems to have had a relatively good view of the events. She is upset: "Go and get him! He is wearing blue jeans and a black t-shirt!" Later, at a press conference, a police officer recounts this description: "A woman saw the murderer, so we know something about how he was dressed. She said that he was wearing blue jeans and a black t-shirt". A journalist then writes an article in the next day's newspaper. After having described the murderer's escape route, the journalist reports on his outfit: "The police said that, according to a witness, he was wearing blue jeans and a black t-shirt". The newspaper gets distributed to thousands and thousands of readers, who may be studying the article in all sorts of different situations: at the breakfast table, at the beach, in a New York City coffee shop, on the back of a camel in the Sahara desert, and so on and so forth. Despite the plethora of contextual variation involved, the content of the witness's description seems to remain constant throughout these transmissions.

Now, it is not clear that Bridges or McDowell would really want to use this point against radical contextualists, and neither Bridges nor McDowell is my target in this chapter. In fact, as I suggest at the end of the chapter, there might be no substantive disagreement between me and them at all. What I *will* do in what follows is to defend the following two claims:

(i) Even if homophonic reports are a central and common part of established linguistic practice, the idea that they are *always* correct is wrong.
(ii) The fact that such reports *often* work fine is no argument against radical contextualism, rightly understood.

To justify these claims, I shall look in some detail at a couple of philosophers who explicitly claim that radical contextualism ignores both the existence and the importance of the sort of semantic stability across contexts just illustrated—namely, Herman Cappelen and Ernie Lepore.[2] Indeed, according to Cappelen and Lepore, most of the expressions that radical contextualists take to be context-sensitive are not context-sensitive at all. In order to substantiate this claim, they invoke a number of context sensitivity tests. The first of these tests is where their version of the homophonic report principle appears. They call it the "Inter-Contextual Disquotational Indirect Report Test", or "ICDIR test" for short. Here is how they describe it:

> Suppose you suspect, or at least want to ascertain whether, *e* is context sensitive. Take an utterance *u* of a sentence *S* containing *e* in context *C*. Let C' be a context relevantly different from C (i.e., different according to the standards significant to contextualists about *e*). If there's a true disquotational* indirect report of u in C', then that's evidence that S is context insensitive. (To be 'disquotational*' just means you can adjust the semantic

[2] I am not the first one who defends (i) and (ii) against Cappelen and Lepore. Others who do so include Bezuidenhout (2006), Picazo (2022) and Recanati (2006). I think most of Bezuidenhout's, Picazo's and Recanati's arguments are compelling, and some of my own complaints against Cappelen and Lepore reiterate points made in their papers. Still, my own arguments differ from Bezuidenhout's, Picazo's and Recanati's in important respects, including the way in which I raise questions about Cappelen's and Lepore's use of a "story-telling context" and their rejection of context-shifting arguments (see sections "Bezuidenhout's Red Apple", "The Nature of Context-Shifting Arguments", and "What Is a Context?"). For other, more specific ways in which my position differs from Picazo's and Bezuidenhout's, see footnotes 3 and 4.

> values of components of S that are generally recognized as context sensitive, i.e, we just test for the controversial components.) (Cappelen and Lepore 2005, 89)

To illustrate how this works, Cappelen and Lepore ask us to consider an utterance by Rupert on Tuesday of the sentence 'John will go to Paris tomorrow'. If, on Wednesday, Pia reports, 'Rupert said that John will go to Paris tomorrow', her utterance will be false, due to her failure to observe how the word 'tomorrow' is sensitive to tense. So, 'tomorrow' is context-sensitive. According to Cappelen and Lepore, applying this test will reveal that most expressions differ from 'tomorrow' in this respect: they do not pass the test.

According to Cappelen and Lepore, this is no coincidence. For, they argue, if all expressions passed the ICDIR test, it would be inexplicable how communication is possible at all. In what follows, I will argue that this claim is false. Even if all expressions passed the ICDIR test, this would not threaten the possibility of inter-contextual communication. Or, to put the point in relation to Bridges's homophonic report principle: Even if we can always find cases that go against the homophonic report principle, inter-contextual communication is still possible.

My arguments lead up to some reflections on the very notion of "context" that usually figures in discussions of contextualist semantics. I suggest that this notion—or, more precisely, the spatial metaphors that tend to come with it ('in context C1', 'outside context C2', and so on)—fosters a profoundly misleading picture of linguistic practice and inter-contextual communication. Once we get rid of this picture, it becomes possible to understand why there need be no tension between saying that language is a practice and saying that it is radically context-sensitive.

Normal and Special Circumstances of Use

Charles Travis says it is a consequence of his radical contextualism that "homophonically constructed 'that' locutions are no longer automatically apt for specifying what given words said" (Travis 1991, 240). Notice what Travis is *not* saying here. He is not saying that we constantly move in and out of different conversational contexts, nor that it will very often be the case that contexts shift so that homophonic reports of what someone said become incorrect. More generally, radical contextualism is not a view about *how often* we move into and out of relevantly different conversational contexts. It is not a view about the actual frequency with which circumstances of utterance change in ways that matter to what is being said by utterances of one and the same sentence. Rather, the radical contextualist claim is just that such contextual shifts are possible with regard to any linguistic expression.

Consequently, a radical contextualist such as Travis may well hold that the context-sensitivity of some sentences is such that variation in their truth-conditions occurs only under what might reasonably be counted as rather special circumstances. For example, in what can reasonably be described as 'normal' or 'standard' situations, an utterance of 'The Eiffel Tower is made of metal' is made true by the fact that the 7300 tons framework of the famous structure in Paris is made of metal. This is usually the case, for example, when the sentence occurs in a guidebook or an encyclopedia. So, there is a sort of semantic stability here, but one that is fully compatible with any sensible version of radical contextualism.

Thus, Travis's way of putting things is to the point. What the radical contextualist is saying is just that there are *some* conceivable circumstances—be they common, or uncommon, or even purely imaginary—under which the truth-conditions of an utterance of 'The Eiffel Tower is made of metal' would deviate from the normal ones. Hence, as Travis says, a homophonic report *need* not be true. And he might use a story such as the following to prove his point: Jim and Cathy are window-shopping in the town of Bordeaux. Very impressed, they stop outside the window of a *Chocolatier* in which a series of wonderful chocolate replicas of famous structures, including the Eiffel Tower, are on display. Jim says, "Do you really think all these sculptures are made of chocolate?" Cathy responds, "No, the *Guide Michelin* says that the Eiffel Tower is made of metal". Cathy is either joking or being downright stupid. In any case, her report misrepresents the content of the guidebook, for the *Guide* was of course talking about the famous building in Paris.

The same point can be made with regard to the murder example I gave earlier. A radical contextualist does not have to deny that truth-conditions can remain stable throughout the series of transmissions described in that example. Rather, she just has to say that some cases are conceivable in which a homophonic report doesn't work. And such cases don't seem difficult to come up with. Thus, suppose Joan is arranging a birthday party. To amuse the children, she buys a special sort of orange paint that can be sprayed on clothes and then removed by washing the garments in cold water. It is a huge success: all the children spray their outfits with the paint. So, when Peter comes to pick up his daughter, her shirt and trousers are completely covered with orange paint. Walking his girl back home, Peter gets a call from Joan, who is worried: "I just read on the spray bottle that the paint shouldn't be applied on clothes with light colors, since visible traces of orange may still be left after wash". "Oh", says Peter, "don't worry, she is wearing blue jeans and a black t-shirt". After the call, Joan's husband asks her, "Were you talking to Peter? You should ask him if his daughter wants to meet us at the stadium in an hour to watch the soccer game between Holland and Germany. We have a spare ticket. But tell him that she shouldn't dress in any way that may provoke the German fans, they are really incredibly irascible". To this question, it would certainly be mistaken of Joan to respond, "She can go there directly, then; there is no need for her to change. Peter said that she is wearing blue jeans and a black t-shirt".

Bezuidenhout's Red Apple

So, I'm warning against conflating the following two claims:

1. All sentences of a natural language are context-sensitive.
2. It often happens that homophonic reports of utterances in a natural language are incorrect.

It is quite possible to maintain the first claim and yet acknowledge that many sentences are such that most utterances of them exhibit the sort of semantic stability that makes homophonic reports true. Of course, if a sentence is used on a series of occasions, the circumstances of utterance will always differ in *some* ways. But the radical contextualist is free to say that in many cases those differences do not matter, since the sentence in question is not sensitive to those particular variations.

Now, Cappelen and Lepore may seem fully aware of this point. For they make it clear that their ICDIR test does not take as input just any pair of contexts. They say the contexts must be *relevantly* different, and by this they mean: different according to the standards that the contextualist herself deems significant to the expression whose context-sensitivity is being tested.

So, at first sight, their construal of the ICDIR test seems sound and straightforward. First, let the contextualist describe two situations that differ in ways she deems relevant to the sentence S that is being tested. Then test her claim by considering whether a homophonic report, "A said that S", issued in one of the situations, will give a correct account of what was being said by A's utterance of S in the other situation.

But in fact, their application of the test is not that simple. It is true that, in applying the test to expressions that they themselves take to be context-sensitive, such as 'tomorrow', they do follow the straightforward scheme just described. But when they apply the test to expressions that they regard as *in*sensitive to contextual variation, they suddenly complicate the test procedure. Consider how they argue against Anne Bezuidenhout's claim that color predicates are context sensitive. Here is a story that Bezuidenhout uses to vindicate her view:

> We're at a country fair picking through a barrel of assorted apples. My son says, 'Here's a red one', and what he says is true if the apple is indeed red. But what counts as being red in this context? For apples, being red generally means having a red skin, which is different from what we normally mean by calling a watermelon, or a leaf, or a star, or hair, red. But even when it is an apple that is in question, other understandings of what it is to call it 'red' are possible, given suitable circumstances. For instance, suppose now that we are sorting through a barrel of apples to find those that have been afflicted by a horrible fungus disease. This fungus grows out from the core and stains the flesh of the apple red. My son slices each apple open and

> puts the good ones in a cooking pot. The bad ones he hands me. Cutting open an apple he remarks: 'Here's a red one'. What he says is true if the apple has red flesh, even if it also happens to be a Granny Smith apple. (Bezuidenhout 2002, 107)

A straightforward application of the ICDIR test to Bezuidenhout's pair of scenarios would be as follows. Suppose that, in the second sort of situation (in which we're interested in finding the apples afflicted by the fungus disease) someone asked, 'Has anyone sorted through the barrel of apples over there?' and got the answer 'You don't have to do that; John said yesterday that all apples in that barrel are red'. Suppose, further, that the purpose of yesterday's sorting procedure was not to separate the apples afflicted by the fungus disease, but just to separate the red-skinned apples from the green-skinned ones. Indeed, we can even suppose that the disease outbreak was unknown at that time. If so, it seems clear that the disquotational indirect report of what John said yesterday is incorrect. So, it would seem as if 'is red' does pass the ICDIR test.

Cappelen's and Lepore's way of handling Bezuidenhout's example is quite different, however. They instead begin by introducing a *third* context—what they call a "story-telling context":

> [W]e're in a café on 5th street in New York between Avenues A and B; it's one of those hot and muggy New York summer evenings, the lights are out, and we're drinking iced tea (look, we're not the ones who think any of this is relevant; we're just trying to humor Radical Contextualists by providing 'relevant' data). Call our context 5stC. In 5stC we're engaging in one of the radical contextualist's favorite pastimes: we're cooking up thought experiments involving distinct utterances of single sentences under diverse circumstances. (Cappelen and Lepore 2005, 90)

Allegedly, it is within this story-telling context that the relevant indirect reports are to be made and assessed. Cappelen's and Lepore's claim is that when, in this third context, we report what is said by the utterances of 'The apple is red' made in the two sorts of contexts that Bezuidenhout describes—the target contexts, called 'C1' and 'C2' respectively—we report that the same content was expressed on both occasions, namely, *that the apple was red* (full stop). Here is Cappelen and Lepore again:

> We are still on 5th street. We're thinking about what Anne said in C1 and C2. Here's what we think about C1:
>
> (2.1) Anne said that the apple was red.
>
> This is what we think about C2:
>
> (2.2) Anne said that the apple was red.
>
> Come to think of it:
>
> (2.3) Both in C1 and C2, Anne said that the apple was red. (2005, 92)

I will soon try to explain why Cappelen and Lepore set up the ICDIR test in this rather complicated fashion. However, before looking closer at their reason for doing so, I want to dwell a little further on Bezuidenhout's apple example. I will suggest that there is a quite plausible contextualist account of what happens when the complicated version of the test is applied to this example—an account that runs counter to Cappelen's and Lepore's view of what is going on.

To begin with, consider again the following report:

> (3) Rupert said that John will go to Paris tomorrow.

Everyone, including Cappelen and Lepore, agree that the sentence 'John will go to Paris tomorrow' is context-sensitive. It will pass the ICDIR test, even in the more complicated version created by the introduction of a story-telling context. Why? The crucial point is this: No story-telling context will be neutral or meager enough to avoid giving what is reasonably called a *context-specific* determination of the semantic value of 'tomorrow'. By this, I mean: There will always be other contexts in which the same expression will not be assigned the same semantic value. Hence, no matter what the story-telling context is like, you can always find a target context that is relevantly different—a target context such that if (3) were used, in the story-telling context, to report the content of an utterance made in the target-context, it would be straightforwardly false.

Now consider:

(4) Anne said that the apple was red.

According to Cappelen and Lepore, this case is different from (3), in that there is a story-telling context that is neutral or meager enough to avoid determining the content of 'is red' in what I called a context-specific fashion—namely, 5stC. More precisely, Cappelen and Lepore make two claims: (a) when it is being used in the story-telling context, all that the color predicate 'is red' attributes to the object to which it is applied is the property of *being red* (full stop); and (b) whenever the color predicate 'is red' is being used, it attributes the property of *being red* (full stop) to the object to which it is applied. Importantly, Cappelen and Lepore do not deny that in *most* contexts, 'is red' also attributes *other*, *additional* properties to the object to which it is applied. For example, there are contexts in which it also attributes the property of having red skin; and other contexts in which it attributes the property of having red flesh; and so on and so forth. Their claim is that those additional attributions are merely pragmatic effects. Allegedly, they are part of what Cappelen and Lepore call the *speech-act content* of utterances, and do not belong to semantics proper.

Thus, the function of the story-telling context is to find a context in which *nothing* really matters, except bringing the "minimal", "purely semantic" content into view—a context in which the expression has *no* real job to do, so to speak. As Cappelen and Lepore notice, this involves a certain kind of difficulty:

> It takes stage setting to get readers to 'see' or 'notice' the semantic content of an utterance. In most regular-life contexts, the semantic content is not what is focused upon. We're interested in all the extra stuff; that's typically where the juice is. However, in certain particular contexts, such as articles on semantics, the goal is to draw the audience's attention to the semantic content of an utterance, and the audience recognizes that this is the goal. With a little stage setting from us, and some receptiveness from the audience, we almost always succeed. (2005, 207)

So, in Cappelen's and Lepore's story-telling context, as they themselves suggest, the summer evening and the ice tea are mere decorations. The essential thing about this context is that it supposedly detaches the speakers from any practical need they might have to employ the expressions under scrutiny. And this supposedly brings out the pure semantic content of the expressions in question.

The radical contextualist of course thinks that this sort of "stage-setting" is deeply spurious, and that the need for it in fact indicates that the invariant content that Cappelen and Lepore postulate is an idle wheel—a superfluous epicycle added just to save certain misguided *a priori* assumptions about how language must function. But then what, exactly, is it that happens in a story-telling context like the one Cappelen and Lepore are describing? Well, with respect to the particular example under consideration here—the English sentence 'The apple is red'—I think the radical contextualist should simply deny that 5stC (or any other allegedly 'neutral' context of utterance, such as a textbook in semantics or articles in philosophy journals) is meager or neutral enough to avoid giving the color predicate a context-specific content, in the relevant sense of 'context-specific'. To see what I mean, note Bezuidenhout's observation that *generally*—or, as I myself put it earlier, in *normal* situations—'red apple' applies to red-skinned apples. Now nothing in Cappelen's and Lepore's description of 5stC suggests that the situation is anything but normal with respect to 'red apple'. So, it seems reasonable to say that in that story-telling context, despite its alleged 'meagerness', 'red apple' means nothing less than *red-skinned apple*. And similarly in semantic textbooks and philosophy journals. Such 'meager' contexts are not isolated from the rest of our lives, lives in which 'red apple' means red-skinned apple unless some very special circumstances, such as the imagined fungus disease, are present. The upshot is that when Cappelen and Lepore are situated in their New York café, their homophonic report of what Anne said in context C1 (in which apples were sorted according to the color of their skin) is indeed correct, whereas their homophonic report of what she said in context C2 (where apples were sorted according to the color of their flesh) is *false*. In order to correctly report what she said in the latter, non-normal context, they need to say: Anne said that the apple had red flesh. Consequently, even if we complicate the ICDIR test the way they do, the appropriate conclusion is that 'is red' is context-sensitive.[3,4]

The Nature of Context-Shifting Arguments

But maybe I'm ignoring the deeper reason behind Cappelen's and Lepore's way of setting up the test. Applications of the ICDIR test involve what they call *context-shifting arguments*—arguments in which we imagine

[3] Bedzuidenhout herself seems to want to go another way than I suggest here. In a response to Cappelen and Lepore, she writes: "[T]he word 'red' and other color terms like 'brown' and 'green' have been both used and mentioned frequently in this paper. So how did I understand these terms? What were the conditions of application for these terms that were operative in this context? I don't think this question makes much sense, because nowhere in this paper did I use the term 'red' to make a genuine color attribution to any real object, such as a real apple. The topic of this paper was not the color of apples. We were not having a conversation about the color of apples. Rather, the main topic of this paper was contextualism, and other sub-topics included such notions as context-sensitivity, indexicality, reference, meaning, content, context, conditions of application for predicates, etc. So the question as to what counts as red in the context of my paper is moot, as is the question of whether I was using 'red' in a self-defeating way" (Bezuidenhout 2006, 9). If I am right, Bezuidenhout's attempt to insulate her own use of color predicates in her article from the making of 'genuine color attributions' does not work and is unnecessary. Instead, she should freely admit that articles on semantics, including her own, are not separable from the rest of our life with language—a life in which 'red apple' means red-skinned apple unless some special reason is given to interpret the expression otherwise.

[4] Whereas I talk about what expressions are used to say in 'normal' contexts, Picazo spells out an apparently similar point in statistical terms: "Metalinguistic information includes information about which interpretations are more frequent. Suppose that most of the time the expression 'red' is used to describe an apple, it refers to the color of its skin, that is, to certain varieties of apples. People familiar with this statistical knowledge will plausibly think that the stereotypical conversation in which the expression 'red apple' is used is one where 'red' described the apple's skin, and use this information to interpret an out-of-the-blue utterance of the sentence 'I've bought some red apples'. Similarly, they can use their knowledge about the most frequent sense of 'red' when combined with 'apple' in order to interpret reports of the form 'Paul said that he had bought red apples'. These interpretations based on likelihood can be called 'interpretations by default'. When we interpret by default we trust that the present use will be aligned with (most) past uses" (Picazo 2022, 267; notes omitted). My position is importantly different from Picazo's. My claim is that it is integral to the know-how involved in linguistic competence that one is able to decide which interpretation is 'default', not on the basis of statistical studies of language use, but (roughly) because one has an understanding of the life-world in which the use of language has its significance. Unless something special has happened—such as the outbreak of a fungus disease—there is no *point* in using 'red apple' otherwise than as a description of the color of the apple's skin. I say a bit more about this towards the end of the present chapter.

language being used in contexts other than the ones we happen to find ourselves in. As Cappelen and Lepore notice, contextualists are very fond of arguments of this sort. However, they continue, such arguments are notoriously difficult to handle properly. Indeed, they argue that the literature on context sensitivity "is plagued by a blatant misuse of this kind of argument". More precisely:

> The fundamental mistake in the entire contextualist literature is this: To properly engage in this sort of thought experiment a theorist has to locate herself *in a particular context*. To not make the context of the thought experiment an essential variable of the experiment is like trying to measure the speed of objects around you while ignoring your own speed. You can't do it. (Cappelen and Lepore, 2005, 10)

This strongly suggests that Cappelen and Lepore think there are good independent reasons for introducing the story-telling context in setting up the ICDIR test. Indeed, they seem to think that a proper application of the ICDIR test *requires* the introduction of this third context. Or better, they conceive of this not in terms of *introducing* a third context, but of making explicit a set of presumptions that are there anyhow, albeit implicitly. Cappelen and Lepore hope to make the contextualist aware of something that, ironically enough, she is prone to overlook: her own communicative situation.

More precisely, their idea is that it is only because the radical contextualist overlooks her own communicative situation that it can seem plausible to her that there are no context-insensitive words or sentences. While engaging in context-shifting arguments, the radical contextualist does not hesitate to specify what is being said in the various target contexts that she is discussing. Thus she claims that in a certain context, 'This apple is red' is true if and only if the apple in question has red skin, whereas in another context the same sentence is true of a Granny Smith apple afflicted by a fungal disease. She thereby takes it for granted that such reports specify determinate contents that are immediately intelligible to her readers. According to Cappelen and Lepore, this by itself shows that the contextualist takes it for granted that her own words have precisely the sort of context-invariant contents that she takes herself to be arguing against: "R[adical] C[ontextualist] descriptions of what speakers say by uttering sentences in various contexts are incompatible with the Radical Contextualist claim that there's no common core, no one proposition expressed in all these contexts, i.e., incompatible with the claim that there's no semantically minimal proposition expressed" (2005, 132).

If I'm right, Cappelen and Lepore are misguided. It is of course true that when a contextualist such as Bezuidenhout uses the expressions 'red-skinned' and 'red flesh' to specify what she takes to be the content of utterances of 'The apple is red' in different contexts, she presupposes that those expressions have a determinate content in the context in which she is currently situated (i.e., in the context of an article on contextualism). It seems to me, however, that all she needs to presuppose is that those expressions, as she uses them, have what I have called their normal content. And I have explained why this normal content is not to be identified with Cappelen's and Lepore's notion of semantic content. The normal content is not a content that expressions have in any context of utterance. There are special situations in which the content is quite different. Hence, the notion of such a normal content is fully compatible with radical contextualism.[5]

This may still seem unsatisfactory. For didn't I suggest, at the beginning of this chapter, that if radical contextualism is correct, then the content of an utterance can never be made completely explicit? Whereas radical contextualists talk as if they *can* specify contents fairly easily, simply by using expressions such as 'red-skinned' and 'red flesh'! Isn't it obvious that this is incompatible with the view they are advocating?

This is again misconceived. I did not say that if radical contextualism is correct, then the content of an utterance can never be made fully

[5] Importantly, my claim is not that all sentences have a 'normal' content in this sense. Sentences that contain indexicals and demonstratives do not; nor do sentences such as 'John will go to Paris tomorrow'. Another interesting case is the sentence 'Caesar was ready'. It has no normal content in my sense. Hence, I would say, in disagreement with Cappelen and Lepore, that no proposition at all is specified in a homophonic report such as 'John said that Caesar was ready' if made in a 'neutral' story-telling context. For a nice criticism of Cappelen's and Lepore's account of 'is ready', see Recanati 2006. A full account of my notion of normal content would of course have to look much closer into what sort of sentences have a normal content and what sort of sentences do not. This would include discussing in more detail various forms of context-sensitivity, including the precise differences between what Recanati calls "saturation" and "modulation" (cf. Recanati 2012). The present chapter is only a preamble to such a more extensive and deep-going investigation.

explicit. What I did say was that no matter how many elements we add to a sentence in order to adjust for indexicality, ambiguity, ellipsis, and so on, we will never reach a point at which the result is context-insensitive. This is a quite different claim. For it is perfectly compatible with acknowledging that, *as used within a particular context*, a sentence can make its content perfectly explicit. What is true is that misunderstandings can always arise, and that if such misunderstandings arise we might want to add material to the sentence in order to successfully convey our message. But such increased completeness of detail is a way of getting rid of *actual* misunderstandings, not of every possible misunderstanding. In the case when our listener is at home in the situation and adequately grasps what we are saying, the sentence we use constitutes a fully explicit expression of its content, in the only sense of 'fully explicit' that we can coherently wish for.

What radical contextualists do take for granted is that they and their readers share an understanding of the expressions employed. This understanding does not involve a grasp of context-insensitive semantic contents, but an understanding of how language and situations are interwoven; a shared situatedness as language users, if that is a helpful characterization. Radical contextualists think there is no substitute for such shared understanding, even in principle. The very idea of a speaker who grasps only what Cappelen and Lepore would classify as semantic contents and with whom we could still communicate if we only had the time and patience to eliminate context-sensitivity from the sentences we use is empty.

Cappelen and Lepore say that radical contextualists naively think they can avoid making the context of their thought experiments an essential variable of the experiment. From a radical contextualist viewpoint, this seems like the very opposite of the truth. Instead, it is Cappelen and Lepore who are being naive here. For they think that, for theoretical purposes, the presupposition that we share a life with language and an understanding of the world in which this life takes place can be eliminated in favor of the analog of an absolute framework of measurement: a completely "neutral" story-telling context in which nothing but semantic content gets expressed by the words people use.

What Is a Context?

Underlying Cappelen's and Lepore's reasoning is the worry that if all expressions were context-sensitive, then intercontextual communication would be inexplicable. "The common thread that runs throughout our criticism of contextualism", they write, "is that it fails to account for how we communicate *across contexts*" (2005, 123).

But is this really true? Consider again the case in which a horrible fungus disease that stains the flesh of apples red has broken out in our apple orchard, and we want to find the apples afflicted by the disease. Someone asks, 'Has anyone sorted through the barrel of apples over there?'. He gets the answer, 'You don't have to do that. John said yesterday that all apples in that barrel are red'. The purpose of yesterday's sorting procedure, however, was not to separate the apples afflicted by the disease, but to separate the red-skinned apples from the green-skinned ones. If so, it seems clear that the disquotational indirect report of what John said was false. However, such failures of communication can be amended. We simply make the report more specific. Suppose the report is answered in the following manner: 'I thought nobody knew about the disease yesterday. Are you sure John said that the apples had red flesh? Wasn't he talking about the color of their skin?' And suppose the reporter then concedes, 'Oh, you're right: he was of course talking about the color of their skin. Sorry, I didn't think about that.'

Corrections of this sort are a perfectly ordinary phenomenon (cf. Picazo 2022, 268–9). The ease with which they are made of course depends on the fact that the contexts involved are different only with respect to a tiny fragment of our language. With regard to the bulk of expressions, including 'red-skinned apple' and 'apple with red flesh', the contexts are not relevantly different. We can easily see that if John had thought he needed to be more specific, he would have used the expression 'red-skinned' instead of just 'red'.

Imagine a variation of the scenario. John has been working all day sorting apples, putting red-skinned ones in one barrel and green-skinned ones in another. And he has marked the barrels 'Red apples' and 'Green apples'. Just before leaving, he discovers that the horrible fungus disease has broken out. It strikes him that tomorrow, when everyone will be focused on the outbreak, his markings will perhaps be misunderstood. So, he re-labels the barrels, writing instead 'Red-skinned apples' and 'Green-skinned apples'. Communication succeeds, and no misunderstanding of the sort described above arises.

This should make us think more about the very notion of 'context' and its employment in debates of the sort I have been rehearsing in this chapter. I am thinking, in particular, of the use of spatial metaphors. A sentence is said to be uttered 'in' a context, or be taken 'out of' context; two speakers may be 'in different' contexts, and hence 'outside' each other's contexts. Such talk can make it hard to resist a deeply confusing picture of linguistic practice as a sort of archipelago, where different contexts are like separate islands. According to this picture, changing contexts is like traveling between two such separate locations, and communicating across contexts is like sending a message in a bottle. Given this picture, radical contextualism will of course seem untenable, for how could the receivers understand such cross-contextual messages unless words carried with them a determinate context-independent content?

But of course, linguistic practice is not such an archipelago, and contexts are nothing like separate islands. The picture is misleading because it makes you forget about the enormous amount of continuity that normally exists between two occasions of utterance in which one and the same English sentence is used to say different things. Indeed, in comparison with this continuity, the relevant differences are usually very slight. We are still on the same landmass, so to speak; all that has happened are some minor changes in the terrain.[6]

In the manuscript from which I quoted at the beginning of this chapter, Bridges writes:

[6] This is not to deny that there are cases in which there are wider and deeper differences, such as the sort of situations Thomas Kuhn describes in his discussions of paradigm shifts (Kuhn 1962), or those that Alasdair MacIntyre describes as "frontier" or "boundary" situations in which a person inhabits two socially and institutionally very different sets of circumstances and thereby comes to feel the strong attraction of a certain sort of relativism (MacIntyre 1985). My point is just that such Kuhnian or MacIntyrean cases are *special*; most everyday instances of communication, including those of homophonic reporting in everyday settings, are precisely *not* like that.

> [T]he use and the point that bear upon the content of an utterance are potentially rich and open-ended enough that there is no guarantee we will have made provision for them unless we conceive the context of utterance as encompassing very wide swaths of discourse indeed—perhaps even the whole of language in use. (p. 47)

He goes on by quoting Wittgenstein, from *Zettel*: "What determines our judgment, our concepts and reactions, is not what one man is doing now, an individual action, but the whole hurly-burly of human actions, the background against which we see any action" (*Z*, 1967, §567). As I understand him, Bridges is here rightly reacting against an archipelago conception of linguistic context. I think he and I would agree that learning to speak means being initiated into a whole life with language; an initiation which includes achieving the ability to understand and orient oneself in the world of that life—its life-world, if you like. To get a better sense of what this might mean, consider the following wonderful example of John Haugeland's:

> (5) I left my raincoat in the bathtub since it was still wet. (Haugeland 1979, 621)

We understand immediately that what is said to be wet here is the raincoat rather than the bathtub. Perhaps we don't even think about the surface-grammatical ambiguity of the pronoun 'it'. This is because we see a point in leaving a wet raincoat in a bathtub, but no point in leaving a dry raincoat in a wet bathtub. This point is nothing mysterious: we don't want puddles of water on the hall floor, we don't want other clothes on the rack to become wet, and so on and so forth. I can write the sentence (5) on the blackboard as an example during a lesson in semantics and trust my students to correctly understand the reference of 'it', since they and I share the life-world in which the sentence would be used. Similarly, (5) can be used as an example in a philosophy article whose readers can be trusted to grasp the reference correctly (indeed, I have just used it in such a way). And yet, it is of course possible to imagine a special context in which the same English sentence would be used to say that the bathtub is wet—say, a future scenario in which raincoats are made of some special material which is very useful in certain ways, but which also has to be stored at a humid place in order to keep its useful properties.

And similarly with what I have called the 'normal' meanings of expressions such as 'red apple'. Our life with language is one in which classifying apples according to the natural color of their skin makes good overall sense. There are many humdrum points in doing so: the skin color is correlated with other important properties that apples have; we can identify the color by just looking at the apples (unlike if we were to classify them according to the color of their flesh, in which case we would have to peel them or cut them open, which would make them deteriorate much faster); and so on and so forth. These points are not context specific in the way the reference of indexicals and demonstratives are context specific. Rather, these points continue to matter even in situations in which we are not talking about any real apples, and have no real interest in their would-be color—situations including articles on semantics, classes in philosophy, and Cappelen's and Lepore's story-telling contexts. Such situations are also part of "the whole hurly-burly" of our life with language, and our apple-talk in those situations derives its significance from our overall way of relating to apples *in that life*. The idea that our employment of words in those situations could be withdrawn from that life and still express some content—be it only a 'minimal' one—is an illusion.

References

Bezuidenhout, A. 2002. Truth-conditional Pragmatics. *Philosophical Perspectives* 16: 105–134.

———. 2006. The Coherence of Contextualism. *Mind and Language* 21: 1–10.

Bridges, J. MS. Pulling Out Contextualism's Roots.

Cappelen, H., and E. Lepore. 2005. *Insensitive Semantics: A Defense of Semantic Minimalism and Speech Act Pluralism*. Oxford: Blackwell.

Haugeland, J. 1979. Understanding Natural Language. *Journal of Philosophy* 76: 619–632.

Kuhn, T. 1962. *The Structure of Scientific Revolutions*. Chicago: University of Chicago Press.

MacIntyre, A. 1985. Relativism, Power and Philosophy. *Proceedings and Addresses of the American Philosophical Association* 59: 5–22.
McDowell, J. 2002. Gadamer and Davidson on Understanding and Relativism. In *Gadamer's Century: Essays in Honor of Hans-Georg Gadamer*, ed. J.E. Malpas, U. von Arnswald, and J. Kertscher, 173–193. Cambridge, MA: MIT Press.
Picazo, C. 2022. Homophonic Reports and Gradual Communication. *Pacific Philosophical Quarterly* 103: 259–279.
Recanati, F. 2006. Crazy Minimalism. *Mind and Language* 21: 21–30.
———. 2012. Pragmatic Enrichment. In *Routledge Companion to Philosophy of Language*, ed. G. Russell and D.G. Fara, 67–78. London: Routledge.
Travis, C. 1989. *The Uses of Sense: Wittgenstein's Philosophy of Language*. Oxford: Oxford University Press.
———. 1991. Annals of Analysis. *Mind* 100: 237–264.
Wittgenstein, L. 1967. *Zettel*. London: Blackwell.

Part III

Counting Matters

The Bolshevik Menace of Anthropological Mathematics: Ramsey and Wittgenstein

Cheryl Misak

Introduction

Wittgenstein said (at some point between 1937 and 1944) that 'mathematics is after all an anthropological phenomenon' (1978, 399). This was an about-face from the view expressed in the *Tractatus*, where he had said that mathematics consisted of 'equations' (1922, 6.375). In this chapter, I argue that the shift was caused by Wittgenstein's most important interlocutor, the great Cambridge mathematician, philosopher, and economist, Frank Ramsey, who died in 1930 at the age of 26. Both Ramsey and Wittgenstein came to see in 1929 that anthropological mathematics, or intuitionism, or what Ramsey called a 'human logic', fits better with the pragmatist philosophical picture they were each starting to develop. My argument will be that by 'anthropological', Wittgenstein meant 'human', as Ramsey used the term.

C. Misak (✉)
University of Toronto, Toronto, ON, Canada
e-mail: cheryl.misak@utoronto.ca

K. M. Cahill (ed.), *Wittgenstein on Practice*,
https://doi.org/10.1007/978-3-031-68656-6_7

My reading of Wittgenstein on this matter goes against the grain of the standard interpretation. The standard interpretation takes Wittgenstein's use of 'anthropology', which starts to appear from 1931, to mean 'ethnology', the comparative study of cultures. (See, for instance, Hacker 2013.) It takes his familiarity with Frazer's *Golden Bough* to be central, as well as the work of Malinowski and Spengler. While these thinkers and the idea that human culture and custom is diverse might be standing off to one side as an influence on Wittgenstein's views on mathematics, I will argue that Ramsey's impact is much more direct and traceable. When Wittgenstein uses 'anthropologisch,' he is echoing Ramsey's idea that we must focus on 'human logic' or logic from the agent's perspective. Indeed, Wittgenstein on occasion uses 'ethnologisch' in the early 1940s, suggesting a difference between an ethnological methodological approach and a pragmatist human focus.[1]

Ramsey and the Early Wittgenstein on Mathematics

In the *Tractatus*, Wittgenstein's treatment of mathematics was far from anthropological. Meaningful propositions, in the *Tractatus*, are those that picture the simple world of objects. The statements of logic are exempt from this requirement, as logic does not make any claims about the world. The truths of logic are tautologous. Mathematics, on the other hand, consists of purely syntactic or formal equations. Unlike the tautologies of logic, mathematics fails the requirement of meaningfulness: 'The propositions of mathematics are equations, and therefore pseudo-propositions' (1922, 6.2).[2]

Wittgenstein finished the manuscript that would become the *Tractatus* on the front, during the Great War. Russell and Keynes managed to get

[1] I thank Alois Pichler for this point and for tracking down every occurrence of 'anthropologisch' in Wittgenstein's corpus for me.

[2] See Kremer (2002) for an explanation of how Wittgenstein paradoxically called equations pseudo-propositions while at the same time said that equations show the logic of the world.

the document out of Europe, which was no small task. Even more challenging was finding a translator for this difficult manuscript. The only person who could do it was Frank Ramsey, the precocious Cambridge mathematics undergraduate with a passion for philosophy. He translated the manuscript into English in the winter of 1921–1922. This was before Ramsey met Wittgenstein, but the two of them, with Russell and Ogden, went back and forth over matters large and small before the translation was categorically approved by Wittgenstein.[3]

In the original manuscript, Wittgenstein had written that 'Russell, Whitehead and Frege have not understood the essential of mathematical method, that is, working with equations'.[4] He dropped mention of his opponents in the final version, but his account of mathematics would continue to have its detractors. They would include Ramsey and the Vienna Circle.

During the summer vacation after his final exams, Ramsey wrote a critical notice of *Tractatus* for *Mind*. In September 1923, while it was still in proof, he went to see Wittgenstein in Puchberg to talk about the book, five hours a day, over a fortnight. At that time, Ramsey was rather in awe of Wittgenstein. Six months later, he returned to Austria for an extended stay, both to be psychoanalysed and to spend more time with Wittgenstein. He also met with members of the Vienna Circle.

On this second trip, Ramsey's awe of Wittgenstein dimmed. After one visit, he wrote home:

> I stayed a night at Puchberg last weekend. Wittgenstein seemed to me tired, though not ill; but it isn't really any good talking to him about work, he won't listen. If you suggest a question, he won't listen to your answer but starts thinking of one himself. And it is such hard work for him like pushing something too heavy uphill. (Misak 2020, 172)

After the next, he wrote: Wittgenstein seemed more cheerful ... But he is no good for my work (Misak 2020, 172).

---

[3] See Misak (2020, in press) for the full story.

[4] von Wright (1973, 44).

During this six-month sojourn in Vienna, Ramsey wrote his undergraduate thesis. It was published in 1925 as 'The Foundations of Mathematics'. He agreed with the logicists Russell and Frege that mathematics is a part of logic and tried to repair a defect in their system. Against the formalist view of Hilbert, he said that mathematics does not consist of 'meaningless formulae to be manipulated according to certain arbitrary rules'. Whatever you think of formalism as an account of mathematical propositions, such as '2+2=4', it is 'hopeless' as a theory of mathematical concepts. The concept <2> occurs 'in everyday life', as in 'It is 2 miles to the station', and it has as much 'ordinary meaning' as any other concept (1925 [1990], 165). Notice that even while defending logicism, Ramsey was already concerned with the meaning of mathematical concepts as they are used by ordinary folk.

Ramsey was also set against the intuitionists L.E.J. Brouwer and Hermann Weyl, who held that mathematics is not the discovery of objective or independently existing entities and facts. Numbers and other mathematical entities are human constructions: a putative mathematical object or fact does not exist unless we can see our way to a procedure for constructing it. A mathematical proposition is true if we could prove it, false if we could disprove it, and neither true nor false if it could not be proved or disproved. Intuitionists thus reject the principle of excluded middle—that every declarative proposition is either true or false. Ramsey asserted that in restricting themselves to what can in principle be constructed by human mathematicians, the intuitionists give up 'many of the most fruitful methods of modern analysis, for no reason, as it seems to me, except that the methods fail to conform to their private prejudices' (1925 [1990], 166).

But neither did Wittgenstein get mathematics right. Ramsey declared his position 'obviously a ridiculously narrow view of mathematics', as it is confined to simple arithmetic (1925 [1990], 180). Ramsey suggested, rather, that Wittgenstein's novel account of logic ought to be extended to mathematics. In an *Encyclopedia Britannica* entry on the foundations of mathematics, Ramsey asserted that mathematics 'consists entirely of tautologies in Wittgenstein's sense' (1929, 831). Wittgenstein did not take this to be a friendly amendment.

The Vienna Circle's Perspective

Ramsey sent Schlick a copy 'The Foundations of Mathematics'. The Circle talked about the paper for two weeks in January 1927 and then intermittently right through to 1929. One of the first entries in Carnap's diary about Ramsey says: 'We talked about Wittgenstein and Ramsey. Very interesting'.[5]

What they found most interesting was Ramsey's extension of Wittgenstein's idea of tautology. Like all empiricists, the members of the Vienna Circle struggled to say why the statements of logic and mathematics are legitimate. They can't be verified by experience, and hence don't satisfy the Circle's observational criterion of meaningfulness. The Circle was happy to find their answer in two moves, the first Wittgenstein's, the second Ramsey's. Wittgenstein argued that the propositions of logic are tautologies, making no claims about the world and hence exempt from the verifiability criterion. Ramsey then argued the same for mathematics.

As far as the Circle was concerned, the two moves taken together were a fundamental turning point in philosophy.[6] Hahn was especially clear that the tautological character of mathematics is absolutely essential: 'If this position can be made out … the existence of mathematics is then also compatible with the empiricist position' (Hahn (1980 [1931], 34).

Ramsey and Wittgenstein on Mathematics 1925–1929

After Ramsey's 1924 Vienna stay, his relationship with Wittgenstein went downhill. At first all was fine, with Ramsey writing warmly to Wittgenstein, encouraging him return to Cambridge and to philosophy. But in the summer of 1925, Wittgenstein made a brief visit to Cambridge, which did not go well. The two had an argument over the value of Freud, after which Wittgenstein cut Ramsey dead. The only point of contact they had

[5] ASP/RC-025-72-06: 42-01-1927 bis 28-01-1927/769.

[6] See Hahn et al. (2012 [1929]).

from then till Wittgenstein returned to Cambridge for good in 1929 was a dispute by letter. It was a dispute over the nature of mathematics.

On June 20, 1927, Wittgenstein had one of his meetings with members of the Vienna Circle—Carnap, Schlick and Waismann—at Schlick's house in Vienna, where they were discussing 'The Foundations of Mathematics'. It seems to have been Wittgenstein's first encounter with Ramsey's paper. During the meeting, Wittgenstein registered an objection to Ramsey's account of identity or '='. It was the first time Carnap met Wittgenstein, and he wrote in his diary afterwards that Wittgenstein was very interesting and original. But Carnap thought Wittgenstein's objections to Ramsey were such that Wittgenstein took a rapid or impulsive position and then tried to find arguments for that assessment. One week later, on the 27th, they all met again, this time at Carnap's house. At this meeting, Wittgenstein dictated a letter to Schlick, for delivery to Ramsey. Carnap typed it up later that evening. Wittgenstein then wrote the opening and closing paragraphs by hand and made revisions to the text he had dictated. He addressed it with a chilly 'Dear Mr. Ramsey' and asked Mr. Ramsey to send a response to his objection not directly to him, but via Schlick (McGuinness 2012, 158–61). Ramsey, after thinking that he might try to break the awkwardness by replying to Wittgenstein directly, eventually thought better of it and did as instructed.

The dispute is hard to reconstruct. Wittgenstein took an identity statement to be part of mathematics, and hence part of those 'equations' and 'therefore pseudo-propositions', and he thought that the symbol '=' could be eliminated without loss. Ramsey's suggestion about identity was designed to shore up the defect he had detected in Russell's logicism, a defect that was only amplified if one tried to eliminate identity, as Wittgenstein did. All I want to draw from this debate is that the one point of philosophical engagement between Ramsey and Wittgenstein between 1925 and 1929 was a dispute about the nature of mathematics.

Ramsey 1925–1929

Ramsey missed the stimulus of his friend after the rift. But he was right that Wittgenstein had been no good for his work. 1926 was a turning point for Ramsey. He was now a mathematics don at King's College,

Cambridge, with philosophy very much at the centre of his interests. He wrote two important papers that year, 'Facts and Propositions' and 'Truth and Probability', in which he started to set out his own account of how true beliefs are related to reality. Although still very much engaged with the *Tractatus*, Ramsey's account ran in the opposite direction. In his 1923 critical notice of the *Tractatus*, Ramsey had registered a number of worries. One was that Wittgenstein's picture theory was good only for elementary propositions, not for the whole range meaningful and important beliefs: 'it is evident that, to say the least, this definition is very incomplete; it can be applied literally only in one case, that of the completely analysed elementary proposition' (Ramsey 1923, 469). If we require the elements of elementary propositions to picture or mirror elements in a human-independent reality, we are left with important kinds of statements—general propositions, causal statements, philosophical and ethical statements—declared nonsense or left as mysterious as to how they fit into the picture theory of meaning and truth.

Ramsey was interested in the full range of human belief. He put forward in these 1926 papers a pragmatist account of belief as a habit or disposition to behave and argued that 'the importance of beliefs and disbeliefs lies not in their intrinsic nature, but in their causal properties, i.e. their causes and more especially their effects' (Ramsey (1927 [1990]), 44). He ended 'Facts and Propositions' with this:

> In conclusion, I must emphasize my indebtedness to Mr. Wittgenstein, from whom my view of logic is derived. Everything that I have said is due to him, except the parts which have a pragmatist tendency, which seem to me to be needed in order to fill up a gap in his system. (1927 [1990]), 51)

That gap was a canyon. In 'Truth and Probability', Ramsey was explicit about his debt to the founder of American pragmatism, C.S. Peirce, and committed himself to a central tenet of pragmatism:

> We have ... to consider the human mind and what is the most we can ask of it. The human mind works essentially according to general rules or habits .... We can therefore state the problem of the ideal as 'What habits ... would it be best for the human mind to have?' (Ramsey 1926a [1990], 90)

The problem of the ideal (the problem of ideal belief and assertion, which for Ramsey is the important part of the theory of truth) is linked to usefulness and to the most we can ask of the human mind. Some beliefs or habits are a better basis for human action than others. Unlike Wittgenstein and the Vienna Circle circa 1926, Ramsey did not seek the holy grail of infallibility. He held that logic must reach for what is 'humanly speaking' right. Such a 'human logic or the logic of truth', which tells us how we should think, 'is not merely independent of but sometimes actually incompatible with formal logic' (Ramsey 1926a [1990], 87). Formal logic is the logic of consistency, but consistency is not enough for rationality. We also require habits of mind (general propositions, inductive conclusions, causal laws) that reliably link our beliefs to the facts. They too are reasonable or unreasonable, 'cognitive attitudes' that, which perhaps not true or false, are still evaluable in terms of reliability (Ramsey 1929a [1990], 147).

Anthropological vs. Theological Mathematics: 1928

After the 1925 publication of 'The Foundations of Mathematics', Ramsey committed himself to writing more about the topic. Ogden, who published the *Tractatus* and helped with its translation (taking full credit for it) had commissioned from him a less technical book on the foundations of mathematics. The *Encyclopedia Britannica* had also commissioned entries on the topic from him. But Ramsey was dissatisfied with the position he articulated in his undergraduate thesis as soon as it rolled off the press. In 1926, he spoke of the 'obstinate difficulties' he encountered when thinking about the foundations of mathematics. One of the difficulties he was assessing was Brouwer's denial of the law of excluded middle. Since there is no proof that A or $\neg A$ holds for all of mathematics, intuitionists prefer not to assume the law. In 'The Foundations of Mathematics', Ramsey had cheekily called intuitionism 'the Bolshevik menace of Brouwer and Weyl' (1925 [1990], 219). To think the ability of human beings to construct a proof had anything to do with mathematical truth or falsity was too revolutionary an idea for him. But two years later, in 'Mathematical Logic', he expressed his disapproval of intuitionism more mildly: 'I cannot persuade myself that I do not know for certain that the "law of excluded middle" is true' (Ramsey 1926b, 229). One can feel a wobble in the double negative. Russell perceived it, noting after Ramsey's death that 'there is less certainty' in 'Mathematical Logic' regarding the rejection of intuitionism (Russell 1931, 478). Perhaps Ramsey saw that his stance on the nature of mathematics did not fit nicely with his emerging pragmatism and human logic.

In 1928, he was still exercised about Brouwer and Weyl. He said that they propose

> to abandon many generally accepted parts of mathematics, and to retain only such propositions as they can prove without using the Law of Excluded Middle ... They think that it is wrong to say that there is a number with a certain property ... unless we have a definite construction for finding one. (Ramsey 1929, 84)

And he was still dismissive about formalists like Hilbert. They

> hope to put an end to this disastrous scepticism by taking an altogether different view of what mathematics is. They regard it as merely the manipulation of meaningless symbols according to fixed rules. (Ramsey 1929, 84)

Ramsey thought that a kind of scepticism loomed for formalists. It seems that we could play any one of a number of meaningless games, and so what passes as mathematics is no better than some other system. As Harold Jeffreys later put it, 'this seemed to me, and to him [Ramsey], that this is saying that it doesn't matter what you do with the bath water provided you make sure that the baby is thrown out'.[7]

In the 1928 'On a Problem of Formal Logic' Ramsey tried to solve the *Entscheidungsproblem,* or the decision problem, that had been reinvigorated by Hilbert at the 1928 International Conference of Mathematicians in Bologna. Hilbert had posed a question of any system of logic, such as that of *Principia Mathematica.* Can we find a procedure or an algorithm

[7] TFL MS/COLL/735: 1/1.

to decide whether any well-formed statement in the system is a theorem or not? Hilbert thought that we *would* find a decision procedure—a method of giving an answer 'yes' or 'no' to whether a particular mathematical statement can be proven. In his paper, Ramsey cited his friend Harold Langford's results and noted that Behmann, Bernays and Schönfinkel had solved other parts of the problem. Ramsey solved another. He also suggested that the answer was to be found in Wittgenstein's characterization of logical truth as tautology. His *Encyclopedia Britannica* entry on the Foundations of Mathematics ended abruptly with that very statement. Ramsey was still making the assertion, not shared by Wittgenstein, that we can extend the notion of a truth-functional tautology to all of mathematics.

But Ramsey's position about the foundations of mathematics was in fact unsettled. Perhaps he was perturbed about the kind of problem Carnap identified in a 1931 paper published in *Erkenntnis*, after Ramsey's death. He said of Ramsey's attempt to save logicism in 'The Foundations of Mathematics': 'This happy result is certainly tempting', but 'we should not let ourselves be seduced by it'. It smacks too much of 'a platonic realm of ideas which exist in themselves independently of *if* and *how* finite human beings are able to think them'. Intuitionism has been called 'anthropological mathematics', but Ramsey's logicism might well be called 'theological mathematics' (Carnap 1983 [1931], 50). Ramsey resolved the issues of *Principia*, but at a high cost.[8]

The expression 'anthropological mathematics was in wide circulation in the late 1920s and early 1930s. In a bibliography on the foundations of mathematics published in *Erkenntnis* (1931, 151), a paper by Oskar Becker is listed, with the title 'Über den sog. "Anthropoöogismus" in der

[8] The context in which Carnap wrote was that Waismann had employed one of Ramsey's moves in 'The Foundations of Mathematics', arguing the axiom of reducibility cannot be a logical principle because it is not a tautology. Carnap, in his 1929 *Abriss der Logistik*, also had endorsed Ramsey's solution. But in his 1931 paper, Carnap was sceptical about Ramsey's argument. He said that Ramsey 'courageously' tried to solve Russell's problems by arguing that the circles of the set theoretic paradoxes are harmless, not vicious. On Ramsey's account, Carnap explained, when we say 'the tallest man in the room', the description is fine as it stands, for the person described already exists. The person is simply singled out, not created, by the description. Ramsey, he said, tried the same tactic for properties: the totality of properties already exists. That was the 'theological mathematics'.

Philosophie der Mathematik' ('On the So-called "Anthropologism" in the Philosophy of Mathematics'). As Becker calls it 'so-called anthrop', there must have been other people who called it thus.[9]

We have seen that Ramsey would have been sensitive to Carnap's charge that his account of the foundations of mathematics was theological. He thought we could only talk about human logic, not God's logic. We don't have any evidence that Ramsey knew of Carnap's worry, although since the paper had been read at the September 1930 at Koenigsberg Conference, word of Carnap's concern might have been circulating before Ramsey's death in January of that year. What we can say is that Ramsey was coming to think that logicism was at a dead end, perhaps for the kind of reason Carnap articulated. But unlike Carnap, who was moving towards a kind of formalism, Ramsey was starting to think that intuitionism wasn't such a menace after all.

We also know that Ramsey was bothered by the issue of open general propositions. For Wittgenstein, what can be expressed in the primary language are the natural sciences. A problem then looms for Wittgenstein (and the Vienna Circle). Scientific theories, laws, and causal hypotheses such as *All humans are mortal* and *A deficit in vitamin C causes scurvy* talk about all instances—past, present, and future. They go beyond actual objects in the world, and so they too cannot be expressed in the elementary or primary language. In the *Tractatus*, Wittgenstein treated the general quantifier ∀ ('all') as depending '*palpably* on that of the elementary propositions' (1922, 4.411).[10] A generalization is an infinite conjunction of its instances. To say that All humans are mortal is to say that Bertrand Russell is mortal, and G.E. Moore is mortal, and on and on. To take it to be anything more would be 'superstition': 'Superstition is the belief in the causal nexus' (Wittgenstein 1922, 5.1361). In 1929, in 'General

[9] I thank Christoph Limbeck-Lilineau for pointing this out to me.

[10] Wittgenstein also said that scientific laws are like meshes applied to a surface, giving it form—we can apply a variety of such 'networks' to describe the world (T: 6.341; 6.35). He also said that signs are rule-governed projections of possible situations. (T 3.5). But it is unclear how a mesh or rule can be expressed in the primary language of the *Tractatus*—how they might be broken down so as to hook up to simple objects in the world. What does fit with the picture theory is his idea that the general quantifier depends on the elementary propositions. Wittgenstein would resolve the tension by dropping the picture theory in the late 1930s in favour of thinking about rules and other things can only be expressed in the secondary language.

Propositions and Causality' and in other notes, Ramsey pointed to the fact that an open generalization goes beyond what we can express and hence is not a proposition that can be strictly true or false in Wittgenstein's sense. A universal generalization cannot be the sum of an infinitely large number of singular judgments, as 'the mind cannot attain to any such infinite conception' (Ramsey 1991, 203). In perfect alignment with his idea that a belief is a habit of action, he argued that a general proposition is a rule with which we meet the future.

All this left Ramsey feeling 'without a clear conception of the nature and purpose of mathematics'. He came to the conclusion that

> We cannot … assume that mathematical propositions in general can be made arguments to truth-functions and treated by the propositional calculus, but must examine the question afresh.
>
> Our old conception that the 'propositions' of mathematics expressed each a true judgment has been destroyed and we have as yet nothing to put in its place. (Ramsey 1991, 204)

Ramsey versus Wittgenstein, 1929

Wittgenstein returned to Cambridge in January 1929. Ramsey wrote to Keynes:

> Can I do anything to entertain Wittgenstein? I don't feel at all sure he wants to see me again, but I should love to do anything I can. There are lots of problems I should like his opinion on, though I fear he will find me so intolerably stupid that he won't want to talk about them. … But do let me know of anything I can do, and please give him in my name any invitation that would be tactful and convenient to him or you. And if you could let me know when he is coming to Cambridge as I shouldn't like too many avoidable engagements. (Misak 2020, 346)

Keynes convinced Wittgenstein to set aside his grievance and try to talk to Ramsey, with happy results. Wittgenstein temporarily moved into Ramsey and his wife Lettice's house and even after he was permanently settled elsewhere, he met with Ramsey a few times a week. Lettice reported that Wittgenstein would stride into the third-floor study, not speak for half an hour except for muttering to himself 'I am such a heel', 'I am so horrible', while staring straight through Ramsey, and then eventually get into a huge philosophical wrangle with him. It seems that they both could end up in tears after these disputes.[11] It was an intense year.

Ramsey fired a sustained volley of arguments against Wittgenstein over this period. He was especially vexed by Wittgenstein's distinction between what can be said (elementary propositions) and what can only be shown (religion, ethics, philosophy). One venue for his onslaught was a piece titled 'Philosophy', perhaps intended to be included in a book he was writing. He opened a draft with a reprimand to Wittgenstein:

> Philosophy must be of some use and we must take it seriously; it must clear our thoughts and so our actions, ~~Otherwise it is mere chatter.~~ or else it is a disposition we have to check ... i.e. the chief proposition of philosophy is that philosophy is nonsense. And again we must then take seriously that it is nonsense, and not pretend, as Wittgenstein does, that it is important nonsense![12]

We must avoid the 'absurd position' of the child in the following dialogue:

> 'Say breakfast.' 'Can't.' 'What can't you say?' 'Can't say breakfast.' (1929b [1990]), 6)

His most famous version of this objection to Wittgenstein's saying/showing distinction is a quip made in 'General Propositions and Causality'. Wittgenstein used to go around Cambridge whistling complex operas. Ramsey's zinger was: 'What we can't say, we can't say, and we can't whistle it either' (1929a [1990], 146).

Ramsey's objection was to Wittgenstein's idea that all statements must either mirror states of affairs or be nonsensical, or be nonsensical yet somehow important. Wittgenstein's whole approach was on the wrong track. One method of doing philosophy, 'Ludwig's', is to

[11] TFL MS/COLL/735: 13/7B.

[12] ASP/FPR.1983.01: 006-02-03.

> construct a logic, and do all our philosophical analysis entirely *unselfconsciously*, thinking all the time of the facts and not about our thinking about them, deciding what we mean without any reference to the nature of meanings. (1929b [1990]), 5)

The *Tractatus* tells us that a proposition is a picture of the world and is disconnected to the self whose picture it is. As Wittgenstein accepted, this position leads to solipsism: '*The limits of my language* mean the limits of my world' (1922, 5.6). But Ramsey thought we have to bridge the chasm between ourselves and that world—our philosophy has to tell us how we make claims about the world. Wittgenstein's idea that the subject isn't in the metaphysical world (Ludwig's primary world 'contains no thought') is disastrous. We need to think about the human facts, not try to do the impossible—try to think about the facts somehow abstracted from all human understanding. In an undated note, Ramsey criticized the idea of having a 'first' or absolute philosophy and concluded:

> Our world is therefore a vague one and the precise is a fiction or construction. We cannot use Wittgenstein's notations like '.3 red /.7 blue'; what colour is that? I have no idea. I could only understand it by translating into intelligible terms. (Ramsey 1991, 55)

Wittgenstein's project was to take the propositions of science and ordinary life and exhibiting them 'in a logical system with primitive terms and definitions' (1929b [1990], 1). It results in a philosophy that is of not much use at all:

> The standardization of the colours of beer is not philosophy, but in a sense it is an improvement in notation, and a clarification of thought. (1991, 55)

Ramsey gave a number of examples where Wittgenstein's project fails, cases for which we need some sort of philosophical explication other than a strict definition. One invoked his small daughter: when we try to explain what *Jane's voice* is, we can't define it, since the constituents into which we would analyse it are highly specific sensations for which we have no name.

In addition to concepts too amorphous to analyse, Ramsey argued that sometimes 'nominal definition is inappropriate, and ... what is wanted is an explanation of the use of the symbol' (1929b [1990], 3). We shouldn't seek, with Wittgenstein, a strict logic of our language. Rather, Ramsey argued, we should seek explications that 'show how we intend to use them in future' (1929b [1990], 1). We are frequently 'forced to look not only at the objects we are talking about, but at our own mental states' (1929b [1990], 5):

> We cannot really picture the world as disconnected selves; the selves we know are in the world. What we can't do we can't do and it's no good trying. Philosophy comes from not understanding the logic of our language; but the logic of our language is not what Wittgenstein thought. The pictures we make to ourselves are not pictures of facts. (1991, 51)

Philosophy must not neglect the 'subjective side' (1929b [1990], 6). If it is construed solely as the search for precise definitions or analyses of concepts, too often it will not be true to the complexity of those concepts. It is clear that in 1929, Ramsey was making the point that Sraffa would later press on Wittgenstein—that he should move away from a closed logical-physical system and move towards to an open organic system. While a primary language might be fit for some propositions, the secondary system, full of generality, vagueness, and human inference, is more important.

Ramsey admitted to having once been under the sway of Wittgenstein's conception of philosophy:

> I used to worry myself about the nature of philosophy through excessive scholasticism. I could not see how we could understand a word and not be able to recognize whether a proposed definition of it was or was not correct. I did not realize the vagueness of the whole idea of understanding, the reference it involves to a multitude of performances any of which may fail and require to be restored. (1929b [1990], 1–2)

But now, in 1929, Ramsey was no longer under the sway of the *Tractatus*. He saw that trying to construct an ideal definition in a perfect language was 'treating what is vague as if it were precise and trying to fit it into an exact logical category' (1929b [1990], 7). And he saw that Wittgenstein's various stances on all propositions that go beyond the primary language was problematic. It robs us of the ability to make good sense of general hypotheses and theories, philosophy, and so much more. We need to think through the natures of various kinds of propositions that are not strictly true: the 'multitude of performances' of logic, mathematics, philosophy, ethics, counterfactual conditionals, scientific theories, and so on (1929b [1990], 1–2). That was Ramsey's project in 1929—he started to deliver accounts of how these kinds of statements could be evaluable 'cognitive attitudes' without being strictly true and false in Wittgenstein's primary language.

Wittgenstein 1929

The year 1929–1930 was crucial in Wittgenstein's philosophical development. It was then that he started to turn away from the logical, sparse landscape of the *Tractatus* and gradually became more oriented towards ordinary human language, meaning is use, and the primacy of practice. This shift towards what we now think of as the later Wittgenstein was very clearly made under the influence of Ramsey.

Wittgenstein said harsh things about Ramsey in the early 1930s—for instance, that Ramsey had 'an ugly mind' and that he was a 'bourgeois thinker' (Klagge and Nordmann 2003, 15). And he made an accusation of plagiarism. In 1932, Neurath wrote to Carnap: 'Wittgenstein is again worried about plagiarism. He claims, so I was told that the papers of Ramsey contain Wittgenstein's ideas, noted carefully by Ramsey'.[13] He was prone to accusing others of stealing his ideas—he did the same to Carnap and Waismann. The charge is especially ridiculous with respect to Ramsey, as the ideas they came to share were recorded in Ramsey's writing during the years Wittgenstein had exiled him from Ramsey's life. We can also see from the passages I've marshalled here that in 1929, Ramsey pressed those ideas on Wittgenstein.

[13] VCA: Neurath to Carnap from October 22, 1932, trans. Christoph Limbeck-Lilienau.

By the time the *Philosophical Investigations* was drafted, Wittgenstein could admit his debt. In the preface, he said:

> since I began to occupy myself with philosophy again, sixteen years ago, I could not but recognize grave mistakes in what I set out in that first book. I was helped to realize these mistakes—to a degree which I myself am hardly able to estimate—by the criticism which my ideas encountered from Frank Ramsey, with whom I discussed them in innumerable conversations during the last two years of his life. (2009 [1953], §4)

It may have felt like two years. But Wittgenstein returned to Cambridge on January 17, 1929, and Ramsey died on January 19, 1930. A year, almost to the day.

Wittgenstein was more specific in the text of the book:

> F.P. Ramsey once emphasized in conversation with me that logic was a 'normative science. I do not know exactly what he had in mind, but it was doubtless closely related to what only dawned on me later: namely, that in philosophy we often *compare* the use of words with games and calculi which have fixed rules, but cannot say that someone who is using language *must* be playing such a game.—But if you say that our languages only *approximate* to such calculi you are standing on the very brink of a misunderstanding. For then it may look as if what we were talking about were an *ideal* language. As if our logic were, so to speak, a logic for a vacuum. (2009 [1953]), § 81)

Wittgenstein and Ramsey Turn to Intuitionism, 1929

When Wittgenstein returned to Cambridge in 1929, one of the topics that he and Ramsey discussed was intuitionism. During this year, they would both change their minds about the nature of mathematics, moving towards Bolshevism or anthropological/human mathematics.

In July of 1929, Wittgenstein gave a talk at the Joint Session of the Aristotelian Society and Mind Association, held in Nottingham.[14] He had told the organizers that he would read a paper on logical form, a piece that tinkered with the machinery of the *Tractatus*. At first, Ramsey wasn't certain he'd be able to attend the conference.[15] Wittgenstein wrote to Russell, asking him to attend, as he was sure no one would understand him. He also told Russell that he wasn't going to talk about logical form after all. Since the Joint Session had advertised 'Some Remarks on Logical Form', and since Wittgenstein had sent them the paper in advance as required, that's what they published in the *Proceedings*. Wittgenstein couldn't have been pleased, for by that point he thought the paper problematic. (In 1933 he would call it 'weak'; Wittgenstein 1933, 415.)

Ramsey did in the end go to Nottingham. He chaired Wittgenstein's session.[16] He had chipped away at Wittgenstein's confidence in the Tractarian idea of logical form over the previous six months, picking up where he left off in his Critical Notice. And he had been engaging with Wittgenstein about infinity. That—infinity—is what Wittgenstein talked about in Nottingham.

In the spring of 1928, Wittgenstein had gone to a lecture by Brouwer in Vienna titled 'Mathematik, Wissenschaft und Sprache', and then met with Brouwer. In the lecture, Brouwer gave a philosophical overview of his intuitionism, stressing the role of the ideal mathematician's mind in creating mathematical truth. The lecture seems to have re-ignited Wittgenstein's passion for philosophy and spurred his return to Cambridge. Herbert Feigl had attended the lecture as well and described Wittgenstein's excitement at the café afterwards:

> Suddenly and very volubly Wittgenstein began talking philosophy—and at great length. Perhaps this was the turning point, for ever since that time, 1929, when he moved to Cambridge University Wittgenstein was a philosopher again, and began to exert a tremendous influence. (Feigl 1981, 64)

---

[14] The published report say it was in Northhampton, but it appears to be wrong.

[15] KCA WJHS/85.

[16] See the minutes, Aristotelian Society (1929). In my biography of Ramsey (Misak 2020), I mistakenly said we didn't know whether Ramsey attended this meeting.

One of Wittgenstein's biographers, Ray Monk, says that the paper Wittgenstein delivered in Nottingham has been lost (Monk 1990, 273). Another, Brian McGuinness, suggests that a draft of the talk was likely dictated to Ramsey and can be found in some notes in Ramsey's hand (McGuinness 2006, 24). The notes do seem to be a copy of something Wittgenstein said or wrote.[17] They are very much in Wittgenstein's voice, in German, with Ramsey's interventions in English, and parts of the text appear elsewhere in Wittgenstein's writing. Wittgenstein told F.R. Leavis at midnight on the eve of his departure for Nottingham that he was setting off to lodge a fair copy of his work with Ramsey for safe-keeping (Leavis 1981, 74). Perhaps the notes are the result of a late-night session on that occasion.

One thing is certain. In 1929, Ramsey and Wittgenstein talked a lot about infinity and intuitionism. There is also testimonial evidence. In 1972, Rush Rhees reported on his conversations with Wittgenstein:

> Wittgenstein said often that Ramsey used to say, in discussion, 'I seem to mean something by it' (perhaps some proposition having to do with infinity in mathematics); or perhaps: 'it does mean something to me'. This was almost like speaking of how it looked to him. At any rate, it is not a way of deciding whether the expression you contemplate does mean anything or not. And Wittgenstein's move was always to ask, 'Well, what do you do with it?' To find what it means, consider its application. (Wittgenstein 2015, 54)

This report is four decades after the fact, and it is muddled in an interesting way. Wittgenstein decisively made the move to link meaning and use only after Ramsey's death. It was Ramsey who, in 1929, was saying that if you want to know what something means, consider its application in action. Perhaps Rhees was reading Wittgenstein's later work into his 1929 conversation or perhaps Wittgenstein himself pre-dated his conversion to Ramsey's view. Or perhaps Wittgenstein was in fact already into Ramsey's mindset in 1929, throwing Ramsey's own principles back at him when it came to the nature of infinity.

---

[17] They can be found here: ASP/FPR: 004-23-01.

In any event, in August 1929, Ramsey wrote two long notes which mark his change of mind. He titled them 'Principles of Finitist Mathematics' and 'The Formal Structure of Intuitionist Mathematics'.[18] In them, he actively explored intuitionist mathematics, opting mostly for Weyl's introduction rules and arriving at his own substitution rules. In his book manuscript, also being written at this time, he began to consider whether the law of excluded middle failed to hold of some propositions.

Ramsey's friend, the mathematician Philip Hall, would say that during Ramsey's day intuitionism was not taken seriously in England.[19] But Ramsey and Wittgenstein were together changing that. Their conversations about the infinite were not, it seems, arguments of the sort they had over the *Tractatus* and over pragmatism. The direction was not Ramsey hammering away at Wittgenstein, but the opposite. Ramsey wrote to Schlick on December 10, 1929: 'Wittgenstein (…) is such a great stimulus and has been doing most excellent work, quite destroying my notions on the Foundations of Mathematics.'[20] Although we have seen that Ramsey was unhappy with logicism before Wittgenstein's return to Cambridge, their discussion seems to have cemented his feeling that he had to rethink his stance on the foundations of mathematics. Indeed, he two seem to have been more cooperatively working through this topic, trying to come to a conclusion. Wittgenstein recalled one discussion:

> I said on one occasion that no extensional infinite existed. Ramsey replied, Can't one imagine a man living for ever, i.e. simply never dying, and isn't that extensional infinity? And, to be sure, I can imagine a wheel turning and never stopping. There is a strange difficulty here: it seems to me nonsense to say that there are in a room an infinite number of bodies, as it were by accident. On the other hand I can think in an intentional manner of an infinite law (or an infinite rule) that always produces something new—ad infinitum—but naturally only what a rule can produce, i.e. constructions. (McGuinness 2016, 23)

[18] These can be found here: ASP/FPR.1983.01: 006-06-1; ASP/FPR.1983.01: 006-06-07.

[19] TFL MS/COLL/735: 9/9.

[20] VCA/114-Ram-4.

Wittgenstein thought, when he recalled this conversation in the 1930s, that Ramsey's argument was 'peculiar':

> Let's consider what experience we would regard as a confirmation or proof of the fact that the wheel will never stop spinning. And compare this experience with that which would tell us that the wheel spins for a day, for a year, for ten years, and we shall find it easy to see the difference in the grammar of the assertions '... never comes to rest' and '... comes to rest in 100 years.' Let's think of the kind of evidence we might adduce for the claim that two heavenly bodies will orbit around one another, without end. Or of the Law of Inertia and how it is confirmed. (Wittgenstein 1975, 304)

As we have seen, Ramsey noted in 'General Propositions and Causality' that, indeed, an infinite statement or a law cannot be verified. Nonetheless, we do mean something by the infinite. A statement about the infinite is a rule with which we meet the future. Ramsey was clearly tempted to go over to Bolshevism in 1929—the position that in mathematics we can only go as far as our human intelligence will take us.[21] The way to understand a statement about the infinite, or a law, is that it is a rule with which we meet the future. An inference ticket, as Ryle would later call it.

Russell thought that this was a sudden move. In 1931, he reviewed Braithwaite's edition of Ramsey's posthumously published papers. He expressed bewilderment at Braithwaite's announcement in the introduction that Ramsey had been 'converted to a finitist view' (i.e. to intuitionism). The reader, Russell said, 'is placed in something of a difficulty'. Ramsey must have 'found flaws' in the argument he made in 'The Foundations of Mathematics', 'but we are not told what these flaws were'.[22] But we have seen that the move away from Russell's logicism had been simmering in Ramsey's mind from immediately after his 1925 published defense of it and came to a boil when Wittgenstein engaged him on the matter. Ramsey gradually came to see, I suggest, that intuitionism might be the expression of pragmatism in the philosophy of

---

[21] The relationship between intuitionism and finitism is not straightforward, partly because the extension of the term 'finitism' is open to debate. In Cambridge of the 1920s, the terms were employed as Ramsey does: loosely and often equivalently.

[22] Braithwaite (1931, xii) and Russell (1931, 477).

mathematics. Truth does not go beyond potential human experience. In 1929, Ramsey started to see that perhaps he needed to be as radical in mathematics as he was in philosophy. Wittgenstein seems to have on occasion also perceived that something like intuitionism fit with his emerging picture. In the first of his 1939 lectures on the foundations of mathematics, he said:

> One talks of mathematical discoveries. I shall try again and again to show that what is called a mathematical discovery had much better be called a mathematical invention. (1976, 22)

Mathematicians decide how to extend the use of our terms into new domains. When Turing challenged Wittgenstein on this point, Wittgenstein said, in a clear echo of Ramsey, that Turing was worried about 'introducing Bolshevism into mathematics' (Wittgenstein 1976, 67).

Wittgenstein would never become a card-carrying intuitionist. He said in the same lectures that that intuitionism is 'all bosh' (1976, 237). But he would remain intensely engaged with what he took to be Brouwer's argument that it is a mistake to think of the infinite as a very large finite. And he would come to the view that a rule in mathematics is good if it works well: 'The truth is that counting has proved to pay' (1978, I/§ 4).

Ramsey died before he could settle his thoughts about the foundations of mathematics. But one thing is clear. Ramsey and Wittgenstein in 1929 were debating whether meaning is a matter of picturing the world or something more human, based on and played out in practice. They both worried during this year about the place of mathematics in such a human logic. If we are to understand where Wittgenstein got the idea of mathematics with a human face, we have to understand how it came about during this critical year in his discussions with Ramsey.

Acknowledgement I thank the participants of the Wittgenstein and Practice conference in Bergen, 2022, for helpful comments when I presented a draft of this chapter, as well as Juliet Floyd, Christoph Limbeck-Lillineau, and Alois Pichler for excellent written comments on a subsequent draft.

References

Abbreviations for Unpublished Primary Sources

ASP/FPR.1983.01: Frank Plumpton Ramsey Papers, Archives of Scientific Philosophy, University of Pittsburgh Library System.
ASP/RC.1974.01: Rudolf Carnap Papers, Archives of Scientific Philosophy, University of Pittsburgh Library System.
TFL MS/COLL/735: Laurie Kahn Ramsey Collection, Thomas Fisher Rare Book Library, University of Toronto.
VCA: Vienna Circle Archives, Noord-Hollands Archief.

Other References

Aristotelian Society. 1929. Minutes of the Nineteenth Joint Session of the Aristotelian Society and the Mind Association Held at University College, Northampton, July 12th–15th, 1929. *Proceedings of the Aristotelian Society, New Series* 29 (1928–1929): 390–391.

Braithwaite, Richard. 1931. Editor's Introduction. In *The Foundations of Mathematics and Other Logical Essays*, ed. F.P. Ramsey, ix–xiv. London: Routledge and Kegan Paul.

Carnap, Rudolf. 1983 [1931]. The Logicist Foundations of Mathematics. In *Philosophy of Mathematics: Selected Readings*, 2nd ed. Ed. Paul Benacerraf and Hilary Putnam. Cambridge: Cambridge University Press, 41–52.

Feigl, Herbert. 1981. In *Inquiries and Provocations: Selected Writings 1929–1974*, ed. Robert S. Cohen. Dordrecht, the Netherlands: D. Reidel Publishing Company.

Hacker, P.M.S. 2013. Wittgenstein's Anthropological and Ethnological Approach. In *Wittgenstein: Comparisons and Context*. Oxford: Oxford University Press.

Hahn, Hans. 1980 [1931]. Discussion about the Foundations of Mathematics. In *Empiricism, Logic and Mathematics*, Vienna Circle Collection, ed. Brian McGuinness, vol. 13. Dordrecht: Springer.

Hahn, Hans, Rudolf Carnap, and Otto Neurath. 1973 [1929]. The Scientific Conception of the World: The Vienna Circle. In *Empiricism and Sociology*, ed. M. Neurath and R.S. Cohen, 299–318. Dordrecht: D. Reidel.

Klagge, James, and Alfred Nordmann, eds. 2003. *Ludwig Wittgenstein: Public and Private Occasions*. Lanham, MD: Rowman & Littlefield.

Kremer, M. 2002. Meaning and Mathematics. *Tractatus', Philosophical Investigations* 25 (3): 272–303.

Leavis, F.R. 1981. Memories of Wittgenstein. In *Ludwig Wittgenstein: Personal Recollections*, ed. Rush Rhees, 63–82. Oxford: Basil Blackwell.

McGuinness, Brian. 2006. Wittgenstein and Ramsey. In *Cambridge and Vienna: Frank P. Ramsey and the Vienna Circle*, ed. M.C. Galavotti, 19–28. Dordrecht: Springer.

———, ed. 2012. *Wittgenstein in Cambridge: Letters and Documents 1911–1951*. Oxford: Blackwell.

———, ed. 2016. *Arthur MacIver's Diary: Cambridge (October 1929–March 1930)*, 201–255. Wittgenstein Studien.

Misak, C. 2020. *Frank Ramsey: A Sheer Excess of Powers*. Oxford: Oxford University Press.

———. in press. Wittgenstein in Green I: Ramsey Translates the Tractatus. In *Proceedings of the 2023 International Symposium on Wittgenstein*, ed. A. Pichler.

Monk, Ray. 1990. *Ludwig Wittgenstein: The Duty of Genius*. New York: The Free Press.

Ramsey, Frank. 1923. Critical Notice: '*Tractatus Logico-Philosophicus*, by Ludwig Wittgenstein'. *Mind* 32 (128): 465–478.

———. 1925 [1990]. The Foundations of Mathematics. In *Philosophical Papers*, ed. D.H. Mellor, 164–224. Cambridge: Cambridge University Press.

———. 1926a [1990]. Truth and Probability. In *Philosophical Papers*, ed. D.H. Mellor, 52–95. Cambridge: Cambridge University Press.

———. 1926b. Mathematical Logic. In *Philosophical Papers*, ed. D.H. Mellor, 225–244. Cambridge: Cambridge University Press.

———. 1927 [1990]. Facts and Propositions. In *Philosophical Papers*, ed. D.H. Mellor, 34–51. Cambridge: Cambridge University Press.

———. 1929. The Foundations of Mathematics. In *Encyclopedia Britannica*, 14th ed., Vol. 15, 82–84 London: The Encyclopedia Britannica Company.

———. 1929a [1990]. General Propositions and Causality. In *Philosophical Papers*, ed. D.H. Mellor, 145–164. Cambridge: Cambridge University Press.

———. 1929b [1990]. Philosophy. In *Philosophical Papers*, ed. D.H. Mellor, 1–8. Cambridge: Cambridge University Press.

———. 1991. *Notes on Philosophy, Probability and Mathematics*. Edited by Maria Carla Galavotti. Naples: Bibliopolis.

Russell, Bertrand. 1931. The Foundations of Mathematics and Other Logical Essays, Frank Plumpton Ramsey. *Mind* 40 (160): 476–482.

Waismann, Friedrich. 1928. Die Natur des Reduzibilitätsaxioms. *Monatshefte für Mathematik und Physik* 35: 143–145.

Wittgenstein, Ludwig. 1922. *Tractatus Logico-Philosophicus*. Translated by C.K. Ogden. London: Routledge and Kegan Paul.

———. 1933. Letter to the Editor. *Mind* 42 (167): 415–416.

———. 1975. *Philosophical Remarks*. Edited by R. Rhees and Translated by R. Hargreaves and R. White. Oxford: Basil Blackwell.

———. 1976. *Lectures on the Foundations of Mathematics*. Edited by C. Diamond. Ithaca: Cornell University Press.

———. 1978. *Remarks on the Foundation of Mathematics*. Edited by G.H. von Wright, R. Rhees, and G.E.M. Anscombe. Cambridge MA: The MIT Press.

———. 2009 [1953]. *Philosophical Investigations*. Translated by G.E.M. Anscombe, P.M.S. Hacker, and J. Schulte, Rev. 4th ed. Chichester, West Sussex: Wiley-Blackwell.

———. 2015. Wittgenstein's Philosophical Conversations with Rush Rhees (1939–1950): From the Notes of Rush Rhees. In Gabriel Citron (ed.). *Mind* 124 (493): 1–71.

von Wright, G.H. 1973. *Letters to C. K. Ogden with Comments on the English Translation of the "Tractatus Logico-Philosophicus."* Edited by G. H. von Wright, with an appendix of letters by Frank Plumpton Ramsey. Oxford: Basil Blackwell.

Technik and *Praxis* in Later Wittgenstein

Juliet Floyd

There is no religious denomination in which so much sin has been committed through the misuse of metaphorical expressions as in mathematics.
—MS 106, 58 (1929). (Wittgenstein *1998* (hereafter CV), 3; Wittgenstein *2015*–contains *Nachlass* references, which use the Bergen Nachlass way of referencing manuscripts and typescripts, in this case MS 106, 58)

Introduction

The above-quoted epigraph is an agenda-setting remark from 1929, addressed to Cantor's diagonal argument and the "paradise" Hilbert (1925, 376) said it opens to us. It contains the seeds of Wittgenstein's later criticisms of "puffed up" versions of the proof.[1]

[1] Wittgenstein (1978), *Remarks on the Foundations of Mathematics* (hereafter RFM) II (1938), see Floyd and Mühlhölzer (2020 §§7–8).

J. Floyd (✉)
Boston University, Brookline, MA, USA
e-mail: jfloyd@bu.edu

K. M. Cahill (ed.), *Wittgenstein on Practice*,
https://doi.org/10.1007/978-3-031-68656-6_8

It also anticipates what would become a continuing and increasingly explicit and sophisticated resonance between Wittgenstein's philosophical investigations of logic and mathematics and his "anthropological" remarks on culture and religion, a resonance most intensively and effectively pursued during his composition of *Philosophical Investigations* (hereafter PI) during 1937–1945. This period broke his 1933–1936 hiatus in writing about logic and mathematics; in 1937–1944 he returned to the foundations of mathematics with a vengeance.

As we shall see, this formed part of his work as a philosopher of culture, something that has implications, not only for how we read his remarks on logic and mathematics, but also for how we interpret his rule-following investigations and his responses to Spengler, Sraffa, and Frazer.[2] By 1938, there emerges in Wittgenstein's remarks a newfangled distinction between the notions of "technique" [*Technik*] and *Praxis*: it is not simply the distinction between *Theorie* and *Praxis* that matters to our notion of *following a rule*. *Praxis* has struck many readers as a key concept drawing Wittgenstein's later philosophizing back to forms of life and culture; but it does not sufficiently dislodge the myth of logical rules as frictionless rails, or the myth of language as a machine. The workaday notions of *technique*, *technology* and *technics* [*Technik*] are finer craftsmen, entering for Wittgenstein to more carefully undo these analogies—which are after all his own. They allow for various ways of inventing, comparing, and contrasting what may—or may not (PI §520)—be included in our practices, allowing for *self*-criticism.[3]

Relations between the notions of *practice* and *technique* in Wittgenstein are made complex by the fact that every "technique" occurs within at least one practice, and counts, it seems, *as* a practice—indeed "mastery" of techniques is part of knowing a language (PI §150). Wittgenstein himself in his lectures sometimes equates a particular practice with a technique. So why differentiate the notions?

First, *Technik* has a particular and long-standing philosophical history in the German tradition, especially in Spengler's work, but also Musil's

---

[2] Works on this from which I have learned include von Wright (1983), Cavell (2013), Bouveresse (1991, 2007), Cahill (2011) and Engelmann (2013b, 2016, 2024).

[3] Diamond (1999) on Rorty and Putnam exemplifies well this focus on multifarious techniques.

and Neurath's. Focusing on its role in Wittgenstein helps us better understand, both his responses to Spengler and Frazer and the evolving character of his "anthropological" turn.

Second, attention to the role of *Technik* versus that of *Praxis* helps us better understand Wittgenstein's remarks on logic and rule-following. Following a rule, we read, is a *Praxis* (PI §202), whereas to understand a language is to have "mastered a technique" (PI §199). Does the latter ground or simply repeat Wittgenstein's move to the former? The answer is that mastery of techniques (plural) is crucial to our concept of rule-following as a *Praxis*. There are ways to put words and symbols to "appropriate" uses, and while in any practice we may dispute what these are, appreciation and engagement with the purposiveness of (perhaps alternative) techniques of use relative to evolving forms of life with language are always part and parcel of the *Praxis*.

How did Wittgenstein get here? I shall emphasize the importance of his "anthropological" treatment of the foundations of mathematics 1937–1944, where he first puts the notion of *Technik* to concerted work. Beginning in late 1936 with the drafting of PI emerged not only his mature style of philosophical writing, but also a more self-conscious and probing series of reflections on the notions of *Praxis*, *Regelmässigkeit*, and *Kultur*. *Technik* helped him secure this maturation in his thinking.

By "mature" I shall mean Wittgenstein's philosophy as expressed in PI (all versions 1936–1945[4]) and his remarks on logic and mathematics 1937–1944, published (partially and posthumously) in RFM. RFM I (1937–38) formed part of the "early" version (the *Frühfassung*) of PI (1937–38, hereafter FF), originally placed just after its (current) first 188 sections as an alternative trajectory for PI's rule-following remarks. Wittgenstein's Cambridge lectures 1937–1941 also belong to this period,[5] as do his *1939 Cambridge Lectures on the Foundations of Mathematics* (1989, hereafter LFM) that Turing attended. *Technik* occurs infrequently in his writings before 1938, and when it does he often associates it with "purely technical" calculi. Already in 1930–31 these anticipate his later, more focused work, as they concern both relations between the "technol-

[4] Wittgenstein (2001).

[5] Wittgenstein (2000) (LC), 2017 (WCL).

ogy" of modern mathematical logic and ordinary language (including rule-following issues), and Frazer's scientism, which interprets rituals as pieces of failed "technology".[6] But in 1937 he began to revisit and use *Technik* in a newly self-critical way—as I see it, partly sparked by a need to explore philosophical aspects of Turing's (1936) "machine" analysis of logic.[7] *Technik* appears 115 times in RFM (1937–1944), *Praxis* only 8; 117 times in LFM, *Praxis* only 12; 15 times in PI, *Praxis* only 6. All the PI remarks invoking *Technik* were added to the manuscript later on (1944–1949). I therefore propose that we see the occurrences of *Technik* in PI as a residue and consolidation of his philosophical work of 1937–1944 with *Technik*.

Technik alludes to and reworks a key notion in Spengler's account of modernity, and in particular his idea that mathematical achievements are signature expressions of a culture. Wittgenstein uses this to investigate and set into relief Frege, Russell, Hilbert, Ramsey, and Turing's remarks about mathematical logic as well as Frazer and Spengler. The intensity of remarks such as those in our epigraph may be seen as part of Wittgenstein's response to modernity and to mathematical logic, but it is important that these form part of his responses to *himself* and his own history. He aims to master and place, rather than to defeat, such *Technik*. Of course, there is only so much the notion can deliver: not every living practice contains techniques in the sense of calculating rules (PPF xi §355).

It is not that Wittgenstein makes a simple application of Spengler in any of his remarks. He ignores Spengler's treatment of specific mathematical examples: these are, as Musil pointed out (1921/1978), ridiculous. Even before he read *The Decline of the* West (1923/1926/1928, hereafter DW), Wittgenstein had expressed a sense of the ending of German culture: he did not need Spengler for this. In the *Tractatus* (1921/1981, hereafter TLP), the key concepts of philosophy are *Scheinbegriffe*, as they were not for Spengler. Wittgenstein concerned himself with modern logic and the philosophies of Frege and Russell, not the examples of Spengler, and his notion of *Sprachkritik* is neither Spengler's nor Mauthner's. The *Tractatus* takes the modern idea of a causal nexus of

[6] MS 108 134, 175 (1930), MS 110, 198–9 (1930–31), discussed below.

[7] Floyd 2016, 2017, 2018, 2020b.

natural necessity to be an illusion—whereas such an idea lies close to the heart of Spengler's organic morphological scheme.[8] Wittgenstein was always careful never to predict anything about history *a priori* and distanced himself from any theory of cultural "deterioration".[9] Spengler's narrative is polluted with its *own* culture, a rhapsodic pseudo-Nietzscheanism posing as a tolerant, high-minded view that has overcome its own Western values.

At the same time, the "affection" Musil felt for Spengler's book was shared by Wittgenstein, and it gave expression to an important tendency of the time; as Musil remarked, Spengler's popular "offenses against mathematics, logic and precision" attracted readers precisely through the "overflowing of a lyrical imprecision into the quadrant of reason" (1921/1978, 135f., 149). Wittgenstein shared Spengler's sense—with Kraus, Loos, Musil, and others—of the death of *Kultur* in modernity.[10] The "general tendency" of the age, Wittgenstein said in 1932, has been "to take away possibilities of expression: which is characteristic of [an] age without a culture"; where once there were "great" philosophers, now there are "skillful" ones.[11] Sensitive to the cultural place of the *Technik* of modern logic—particularly after the Vienna Circle's appropriation of the *Tractatus*—Wittgenstein had already given his "Lecture on Ethics" (1929/1965), a public event distancing himself from them. In 1932, he re-emphasized his idea of logic and philosophy as *activities*, telling his students that he aimed to communicate, not a doctrine but a certain kind of critical *technê*, or skill—the only truly philosophical way to master *Technik* in a world without *Kultur*.[12]

It is Spenglerian to treat a culture's *Technik* as culturally indicative for the possibilities of expression. And in this vein Wittgenstein criticized certain metaphors in modern mathematics even while he rejected, as many others did, Spengler's hyperbole and polemics. But he had to find

---

[8] Cahill (2011) and Engelmann (2016).

[9] LC §33.

[10] Engelmann (2023).

[11] Wittgenstein 2014 5:2 (1930).

[12] The resonances between *Technik* and the ancient Greek *technê* are complex and at work in Spengler; compare Heidegger (1993), Schatzberg (2018), Johansen (2021). The notion plays little role in English-speaking philosophy.

a way to do so undogmatically. Neurath (1921/1973) countered Spengler's declinist views by re-casting the modern engineer —who for Spengler does the self-enslaving "Devil's work" of "the Machine" (DW II 5.2.6-7)— as a hero repairing the societal boat of science at sea.[13] The Vienna Circle proposed that philosophy should improve society with science. By contrast, in his mature writings on logic and mathematics Wittgenstein—trained as an engineer himself—*anthropologized* the "beginnings" of mathematics and the image of the machine, rendering *Technik* of logic ordinary or home-baked [*hausgebackene*],[14] that is, subject to human mastery and critique. This miniaturizing dis- and re-assembles parts of what Spengler called the "Machinen-*Technik*" of modernity. It also re-orients and corrects what the *Tractatus* had said about colloquial language being logically in order as it is (5.5563). Words are not reflections of *the* logical structure of sentences, but *tools* for the later Wittgenstein. And there is not just one machine, no "Super-Mechanism" of thought, no pure system of *Technik* (Gödel and Turing's work established this for arithmetic and logic).[15] "Only through a [human mastery of] *Technik* can we grasp a *Regelmässigkeit*" (MS 164 FCv3 (1944)).

In the end, to clarify the whole idea of "use" and meaning and *Praxis*, he *had* to discuss *Technik*:

> One can talk about the function of the word in the sentence, in the language game, in the language. But "function" in each of these cases means "technique" and therefore refers to a general explanation and training. (MS 167, 31 (c. 1947–1948))

This spells out, self-critically, what it is to work with the idea of "meaning as use". The crucial idea—clear only by 1937—is that any given word or sentence or language involves *multiple* techniques of possible use: mastery involves layerings and interweavings of many possible techniques for applying words in situations, and the generality of these is purposive and evolving, rather than quantificational.[16] Wittgenstein reached this point

[13] Soulez (2019, 2023).

[14] MS 157b, 8r, MS 152, 96 (1936–1937); Floyd (2020b).

[15] Floyd (2001, 2016, 2017).

[16] Floyd (1995, 2001); Floyd and Mühlhölzer (2020 §8.2).

while revisiting the "foundations" of logic mathematics: his "anthropological" turn in these areas went hand-in-hand with the development of his later philosophy as a whole, forming part and parcel of his philosophy of culture.

Spengler, Frazer, and Technik

For Spengler humanity's practical "mastery" of Being always involves technics of culture (DW; Spengler 1932). Modern economic life is said to evince a "Faustian *Technik*": humanity's will to power has led to its self-enslavement to *the Machine*.

Technik is exemplified for Spengler by scientific laws, mathematics, religious myths and dogma, and modern economic arrangements (DW I.4.II.x). He emphasizes the recurrent tendency of cultures to decline once technical achievements, for example, cults, begin to exhaust its forces of growth. In *Decline of the West*, we read that

> The means whereby to identify dead forms is Mathematical Law. The means whereby to understand living forms is Analogy. (DW I.1.ii)

As modes of expression, mathematical achievements distill the essence of a culture, but they index the culture's hardening into a mere "civilization". Their forms solidify a rigid manner of *Technik*, a frozen set of procedures that express the beginning of the end of a culture's living progress. Things become merely "technical". After a culture's decline and death, the forces of life, through creative "analogy", begin a new culture, for which a new "culture-language" [*Kultursprache*] is required (I.II.i.xi, I.IX.1.i, II.V.2.vi). According to Spengler, "primitive man" and children lack this, they "*have no Culture* [*Kultur*]" (DW, I.II.i), for culture involves possibilities of (self-)expression. Penetrating metaphors are crucial to implement new growths: true history requires a "*Technik* of analogies" (DW I.1.ii).

Wittgenstein's post-1937 "anthropological" work on the first steps of mathematics aims to bring *Technik* to life, overcoming frozen, rigidified forms, analogies, perspectives—not through inspiration, but by way of mastery and understandings of *Technik*. Language-games embedding, juxtaposing, extending, and developing elementary techniques for the uses of words are a way to resist Spengler's idea that the *technische* achievements of mathematics inevitably signal cultural decline or apotheosis, while granting that they express *possibilities* of culture. This is also a way to respond to Hilbert's view that

> The instrument which mediates between theory and practice, between thought and observation, is mathematics; it builds the connecting bridges, and makes them ever sounder. Thus it happens that our entire modern culture, in so far as it rests on the penetration and utilization of nature, has its foundation in mathematics. (1930, §22, 1162)

By slowing actions down to scrutinize techniques, Wittgenstein aims to explore and fashion new metaphors, investigate what we—even perhaps children, but also mathematicians—are inclined to say, highlighting the significance of the cultural capital we have invested in training ourselves to follow rules and calculate.

For the mature Wittgenstein there is nothing "purely technical", as he had once thought when he remarked that "what is human in thinking has nothing to do with logic" (MS 108, 242 (1930)). His writings on the foundations of mathematics culminate in the maturer idea that mathematics is a "MANY-COLORED MIX of proof techniques" (RFM III §46 (1939))—and he does not say here practices or conventions. His point is that if we include the human element in our conception of logic, we must include the possible mastery of diverse techniques, techniques that may be either discovered or invented. Our mastery cannot be captured once and for all in *fixed* forms of stipulative or abbreviative or formalized "definitions" or systems.[17]

By 1937, Wittgenstein began to systematically investigate the metaphors we are ourselves inclined to use in describing different logical and mathematical achievements, testing them against the particular techniques the metaphors encapsulate. Emphasis on our capacity for *self*-scrutiny earmarks his mature way of responding to Spengler and to Frazer

---

[17] LFM 33–43.

and the mathematical logicians. In this respect, there are certain interesting contrasts we may detect between his earlier and later sets of remarks on Spengler (1930–31, 1933–34 (hereafter RS I), versus 1936–37, 1941–42, 1946 (hereafter RS II)), and his earlier and later remarks on Frazer (1930–31, hereafter RFGB I), versus after 1936 (hereafter RFBG II).[18]

In RS I, Wittgenstein finds in Spengler "many significant thoughts" "in spite of many irresponsibilities in the particulars…"; "[m]uch, perhaps most of it, is completely in touch with what I have often thought myself" (PPO 25). He finds some "good" thoughts in Spengler's way of seeing cultures as organic wholes having a life and death—rather than, in the modern way, seeing history as "a sausage that can go on indefinitely" (PPO 27–29). But he insists that one must not dogmatically confuse, as Spengler does, the motif one uses to describe history with the objects one measures. He suggests that Spengler could be "better understood" if one saw him comparing different cultural periods as one might compare different families—for example, of numbers—along a variety of differing dimensions of resemblance and difference (CV 21).

In RFGB I, Wittgenstein criticizes Frazer by applying to him Spengler's quasi-anthropological aspiration—failed though it was—to avoid imposing one's own understandings upon foreign eras and cultures. He takes Spengler's notion of a *surveyable presentation* [*übersichtliche Darstellung*] to be "of fundamental importance to us" (RFGB I, 133): Such a presentation allows one to see significant connections between apparently different things in a way that an empirical hypothesis about linear evolutionary processes cannot. And he criticizes Frazer's modern prejudice in misdescribing human religious rituals as instruments or "technology" (*Technik*): Frazer fails to distinguish between *Technik* (e.g. the building of huts and the carving of arrows) and religious, symbolical expressions of religion and magic (RFGB I, 125).

[18] RFGB I and II are published in Wittgenstein (1993a) (hereafter PO). The first set of remarks on Spengler occur in 1930 (MS 183, 16,19,29=2003 (hereafter PPO) 25,27,37); 1931 (MS 110, 12–13=CV 12), 257; MS 111, 119–120=CV 21–22 and 195–6=CV 22–23); MS 113, 102–3; 153a 121–2; MS 154, 15v= CV 17 and 1933–34 (MS 115, 56–57); the second in 1936 (MS 142, 107; MS 183, 211=PPO 219), 1937 (MS 157b, 15v= CV 30), 1941–42 (MS 125, 30) and 1946 (MS 130, 239–40=CV 52).

Wittgenstein's later remarks on Spengler and Frazer incorporate a more reflective emphasis on *ourselves*, seeing what *we* can bring *to* others by way of making *associations* among differing practices (PO 143). In RFGB I examples of this had been given (kissing the picture of one's beloved, 123), but Wittgenstein becomes more explicit now that "bringing [a picture of a rite] into connection with our own feelings and thoughts" is what "gives the account its depth" (RFGB II 143). He emphasizes the importance of our developing a *variety* of different ways of entering into others' uses of pictures and words. The capacity we have to see something sinister *in* a child's game or another's ritual may be revealed in considering its origins, not by way of a general hypothesis as in Frazer, but instead by way of a picture from our side of *possible* origins, appealing to activities we recognize *in* human beings generally, a "common spirit" that connects us *with* them (RFBG II, 151). The *aspects* in which these practices appear through our imagined surrounding histories and circumstances express *our* understanding of the internal "character" of the people and the practices involved (RFGB II, 143), showing who we are, allowing for self-criticism and forms of engagement with other cultures. Mastery involves self-mastery.

The notion of *Lebensform*, surprisingly, appears neither in RFGB I nor RFGB II, although the thoughts just sketched hold the seeds of that notion. This *may* suggest that the (still controversial) date of composition of RFGB II is before the fall of 1936, when Wittgenstein first drew that notion into his philosophizing—a deepening move of great importance to his later philosophy.[19]

In any case we can see a similar evolution going on between his earlier and his later remarks on Spengler. In RS II (in 1936), he remarks that "a major source of *our* misunderstanding is that we fail to see how *our* words are used. – *Our* grammar lacks a survey [*Übersichtlichkeit*]" (my emphasis), applying Spengler's idea of a *Weltanschauung* to his *own* way of looking for surveyable presentations of grammar: the aim is to "see connections" and find links between uses of words (MS 142, 107). Noting that he sometimes inserts his remarks "in the wrong place" so that they "don't *work*", he criticizes Spengler's remark that Kierkegaard merely "played"

[19] Floyd (2016 §2).

with language:[20] Kierkegaard was "too big" for him and stands too close", Spengler could only see "the giant's boots", that is, he could not see how Kierkegaard's language *works* (MS 183, 210 (1937)). Asking how Spengler's reflections relate to his newly emerging self-criticisms of his "dogmatism" in the *Tractatus*, he answers that the general form of proposition's role in showing *a priori* "the *nature of the world*" was an "ideal" that now "must disappear" (157b, 3v). Logic will not "lose its rigor", "it just plays a different role now[:]... ... from a prejudice about reality, it has become a form of representation" (157b, 5r). "Spengler's injustice" was to conceive of the world and life according to an *a priori* scheme, but an ideal "loses none of its dignity when it is presented as the principle of the form of representation", that is, *one* way of looking at things (157b, 14v). His answer, then, is that he had *himself* dogmatically made of an object of comparison into an ideal, transforming a "good measure" into a "prejudice" to which everything had to conform (157b, 16v).

He immediately begins sketching language-games where machines, counting, and measuring proceed in a variety of ways, according to a variety of differing techniques and standards of use. The culmination of these remarks is an anthropological agenda, and it will be pursued further in RFM I: to distinguish the role of surprises in

> "Laws of thought" [*Denkgesetze*]
> Habits (or Customs) of thought [*Denkgewohnheiten*]
> Ways of thought [*Denkarten*]. (MS 157b, 37v)

Wittgenstein's "anthropological" agenda with respect to the nature of logic is designed to achieve critical distance on what *we* (and other philosophers) *say* about the necessities of logic and mathematics, which should—precisely to retain its rigor—be living and expressive, organic and not dead or cultish. In 1937, he began to transform *Technik* from a Spenglerian measure of modernity's decline into a living measure of *himself* and his own techniques. Logical necessities turn out to be, not essential contracts or forms that necessarily and eternally bind us, nor natural laws of human behavior we attend to prudentially, but evolving activities

[20] DW II.2.5.iv.

of self-critically embedding words, analogies, and trains of thought in *forms of life*.

Of course, Wittgenstein rejects Spengler's idea of "high Culture". The term *Kultur* never occurs in any version of PI; it is replaced in the fall of 1936 by *Lebensform*, in Wittgenstein's aborted revision of *The Brown Book*. In the Blue and Brown Books, to imagine a language is to imagine a "culture", but in PI it is to imagine a *Lebensform*.[21] These are now what language-games express and reflect in an age "without culture". With its biological and cultural, as well as localized and universal connotations,[22] the mature Wittgenstein's notion of *Lebensform* recapitulates in a new, non-linear, miniaturized, and "gappy" way Spengler's idea that eras in human history involve *possibilities* of (self-) expression and modes of understanding and articulation.[23]

Metamathematics

Wittgenstein's anthropological turn involves confronting the work of, for example, Hilbert, Brouwer, Ramsey, and Turing as expressions of culture. The first occurrence of *Technik* in Wittgenstein's *Nachlass* I know of occurs in 1930, right at the beginning of MS 108. It invokes a particular way of responding to empirical psychology and to metamathematics:

> The methods of the newer mathematical logicians are reminiscent - I believe - of the methods of the current experimental psychologists. In both, something specific is being done, but both are wrong about the meaning [*Bedeutung*] and scope of their work. The tests of intelligence, presence of mind etc. do test something but not what we call presence of mind, intelligence etc., & the proofs of consistency probably show something but nothing important. At least that's what I think. They always crawl around on the surface of the questions and don't see what's actually important. But once the actual work has been done, many of those external games are

[21] Floyd (2016).
[22] Cavell (2013, 42).
[23] Floyd (2020b).

> obsolete or have to be completely reinterpreted. (Again, here is a lot of *Technik* and no spirit [*Geist*].). (MS 108, 134 (1930))

The contrast drawn is between the "spirit" expressed in our "internal" games with psychological and mathematical concepts—games where "we call things" by certain familiar names— and on the other hand the "technics" that grow out of these games into a more "external" way, representing our uses of concepts systematically.

At this time Wittgenstein treated systems of grammar as distinct from one another, so that the distinction between "internal" and "external" features of "games" is sharp, but there is no "metalinguistic" level. According to his line of thought, Hilbert's metamathematical pursuit of consistency proofs for formal systems is not a foundation or analysis or clarification of the meaning and scope of mathematics at a metalevel, but rather just a new form of mathematics with a grammar of its own, similar to the way geometrical axioms offer systematizations of space, and psychological axioms or experiments systemize and analyze the grammar of mental or visual space. Such theories offer particular "technology" [*Technik*], a codification which does not transcend our ordinary life with mental concepts and experiences of meaning but "crawl[s] on the surface", reflecting a manner of presenting conceptual arrangements, not deeper truths or laws.

This retains the *Tractatus*'s idea that there are no *truths* of logic or logical *laws* that can justify our reasoning and calculating, and that everyday language is already logically in order as it is (TLP 5.132, 6.341-2). In logic and mathematics, for example, the generality of step-by-step substitutions into formal series and the making of calculations and solving of equations "takes care of itself" and is always articulable in a particular form that is exact and unambiguous.[24] We "must" grasp the generality of

[24] MS 108, 174—175 (1930). Interestingly the material for the later PI wayward pupil's series appears here. Wittgenstein writes down the series 2,4,6,8 in binary array form and then attempts to express a term for the development of the series, remarking "But this sign must be accompanied by a description or instructions for use. And the difficulty is precisely that of giving it in an exact, that is to say, unambiguous form".

any such procedure in order, for example, even to calculate with decimals. Thus it is a "purely technical" [*rein Technische*] verification to solve for *x* in

$$x^2 = 2x \supset x = 0 \vee x = 2$$

because I can calculate from the equation, inserting via step-by-step substitution the value of *x* "everywhere", and get a true or correct sentence (MS 105, 40 (1929)). Similarly, in metamathematics we can formulate step-by-step rules and induce over and/or systematically substitute letters into formulas to get results. On the one hand, Wittgenstein does not know how to express all these rules exactly: that would require "complicated technology" [*kompliziert Technik*], but on the other hand he "suspects" that "if you only see the internal relation of the members of the form series everything is fine & … there can be no method at all" with which to make substitutions because no method is needed or is possible (MS 108, 175 (1930); compare MS 105, 42 (1929)). From this point of view, particular examples of substitutions or calculations with the rules are only a technical [*technische*] aid, "and once I [have] seen what is wanted, I could also see [the rule] in [any] example, … [so that] it is the rules that apply to the example that make it an example" (MS 111, 94 (1931)), rather than examples providing any foundation for the general rules.

A joke about this drawn from the journal *Simplicissimus* crystallizes these ideas:

> (*Riddles of Technology*. Picture: Two professors in front of a bridge under construction.) Voice from above: "*Laß abi —— hüah —— laß abi sag'i —— nacha dreh'n mer'n anders um!*"
>
> It is unbelievable, Mr. Collegian, that such a complicated and precise work can be done in this language. (MS 111, 132 (1931))

The professors have the ridiculous idea that local dialect is incapable of being used to construct a complicated bridge: it is like Hilbert assuming that his metamathematics offers a foundation for the rest of mathematics and logic. This is to fail to see that the system is already in place in ordinary language if it is there to be elaborated in a theory at all: theory cannot "transcend" it, but only elaborate it.

In the years 1929—1933 Wittgenstein no longer insists that there is *one* (schematic) rule for logic: he developed his *Satzsystem* idea, according to which logical (grammatical) necessities are relative to conventional grammatical systems, of which there are many, each complete and entire.[25] For the Middle Wittgenstein, "two" in the natural numbers signifies a wholly different number from "two" in the rational numbers or the reals, because the rule-systems differ: no incorporation is involved in the development of new number systems from older ones except in terms understandable only from the perspective of the new, broader system itself. The formation of the decimal expansion of a real number thus creates a new grammar at each step of the calculation, and the formulation of unproved mathematical conjectures is impossible, as their grammar is as yet undetermined. In general, Wittgenstein holds that "grammar is arbitrary", that is, autonomous and unjustifiable, and so we cannot compare a sentence in one system of grammar with one in another. And to each sentence of a language that is true or false there is tied a single determinate form of verification.

Wittgenstein's step to language-games in *The Blue and Brown Books* (1969, hereafter BBB, BlB, BrB) 1933–1935 placed this pluralism and conventionalism into a less dogmatic "anthropological" frame, transposing the whole idea of presenting logic in terms of rules and systems. It allowed him to treat the evolution and learning of language and grammar as an organic matter, decentred: rather than the accretion and adoption of absolutely separate systems of rules, language-games display a dovetailing of interpenetrating human uses of words in irreducibly complex conglomerations of use. "Primitive" languages *also* express logic and culture, contra Spengler.

However, in BBB the presentation of language-games is *naively* anthropological, in that the snapshots are trotted out in a linear way, built up step-by-step, adding one kind of word after another, arranged as if in a kind of scheme that moves from what is simple to the more complex: the

[25] Wittgenstein (1980), Floyd (2021) and Engelmann (2013b).

style is "boring and artificial", as he later remarked.[26] This too easily invites the idea of a serial approach to the development of human culture: the imprints of Spengler and Frazer's ways of thinking had left a misleading mark on his thought.

There are nevertheless interesting developments within these dictations to students. In BlB, Wittgenstein emphasizes what happens "in the practice" of language, but only in BrB does he begin to move towards conceiving his language-games as *themselves* "practices" (BrB §32), drawing in the beginnings of reflection on what counts as a practice as such. He imagines a finite table of symbols used as explicit commands to generate "steps" taken by a human being, and then draws in the question of what it is to train a person to follow *any rule as such* ("*general training*", BrB §§38–41, Wittgenstein's emphasis). Now his idea of practice is linked to that which may be presented in terms of rules of operating with signs, and there may be different rule-representations associated with particular practices in language.[27]

In the process of beginning PI, he came to regard this move to "general training" as a problem, one that sparks the need for a general investigation of rule-following and *Regelmässigkeit*. In his initial draft of PI §51 (2001, *Urfassung* §50), investigating the idea of "simple" names with a game involving a table of colored squares, he asks for the first time, so far as I know, "And what does it *mean* to say that *in the practice of language* [*Praxis der Sprache*] the signs correspond to certain elements?" (first emphasis mine). The most important thing he comes to appreciate is that he had been assuming for a long time that each sentence is associated with *one* technique of use, *its* technique of use. But he now has to "get used to the idea" of a *variety* of techniques associated with each sign:

> How is the application of a rule fixed? – Do you mean "logically" fixed? Either by further rules or (not at all). – Or do you think: How is it that everyone agrees to use them so, and not differently? By training, drill, and the forms of our life. This is not a consensus of opinions but of forms of life.

[26] MS 204, 257 (1936); compare Engelmann 2013 §4.4, Floyd (2016, 17).

[27] In the Skinner Archive 1934–1937 (2020) "technique" occurs twice (240,261).

> So I do not say, "Do not believe that the logical laws must be valid everywhere, there may very well be a case in which the law of the excluded middle does not hold" - as if this were an empirical proposition and one had to warn against blindly assuming that he agrees! – But I do say: get used to seeing a diversity [*Mannigfaltigkeit*] of techniques of sign usage (i.e. of thinking). I do not want to eliminate a prejudice of opinion but of technique [or: technology, *Technik*]. For example, do not be dismayed in principle by a contradiction...
>
> ...[b]ecome aware of the uselessness of the ordinary concepts of *Sinn, Bedeutung* and *Wahrheit* ... (MS 160, 27v-r (1938))

In *The Big Typescript* (2005, hereafter BT), he had remarked that the "disastrous" preoccupation with the *Sinn* of a sentence, the "thought" it expresses, had unwittingly driven philosophers to postulate mental images and psychological accompaniments to meaning, whereas "everything depends on the technique for using the sentence", and he "guesse[d]" that "the *sense* of a sentence, in Frege's sense, is its *use*" (BT 267v). His aim there was to sustain and elaborate Frege's logical, anti-psychological conception of thought. But as of 1938 he now sees that a radical revision of a certain kind must come, altering the way he is to handle and incorporate Frege. Because everything depends not on *the* technique of using a sentence, but on the manifold *techniques* for using words, their embedding in forms of life.

Rule-Following

For the mature Wittgenstein, the "mastery" of techniques is held to reflect one's understanding and grasp of a word, sentence or system *as* way of living with language (RFM VI §34). The technics of language may be there, but we *use* them one by one, inventing, discovering, producing, and developing them. PI §199 baldly states that "to understand a language means to have mastered a technique". This responds to the hypothetical question of whether "following a rule" is something just *one* person, only *once* in a lifetime, could do. Wittgenstein answers that, as to the *grammar* of "following a rule", the answer is No: rule-following, making reports, giving orders, playing games of chess are "*customs* [or *habits*] [*Gepflogenheiten*] (usages, institutions)" (compare PI §198; PI §§204,337 nearly equate *Technik* and *Gepflogenheit*). This is not to say that it is empirically or conceptually impossible for a child to invent a game by herself, play once, and then be annihilated, or for an accidental machine to fall from the sky and begin printing digits (cf. RFM VI §§40–43), but rather that such imagined cases, insofar as they involve attributing our general concepts of *rule-following*, *ordering*, *reporting*, and *playing* to the child or the machine involve us in representing them by way of *our* purposive human forms of life, in which particular and various reproducible, structured actions, repetitions, and samenesses and differences of purpose and significance in relation to *us*—to *anyone*—are possible. Our applying our techniques is a contribution to keeping a practice's reality as a practice in play.

The distinction between *Praxis* and *Technik* is thus intimately related to rule-following. The "paradox" remarks (§§201–202) made central by Kripke were inserted in the late version of PI as an interpolation between remarks invoking *Technik* (§§199, 205): they serve to spell out this context. The "paradox" of a regress of interpretations for interpretations, shows, Wittgenstein remarks, that "there must be a way of grasping a rule that is *not* an interpretation", but is instead "exhibited" from case to case in what we call "following the rule" and "going against it" (PI §201). Wittgenstein draws the conclusion in §202 from the regress of interpretations that "*therefore* 'following a rule'" is a practice [*Praxis*]"—which means, at a minimum, rule-following is *Praxis* as opposed to *Theorie* or interpretation. But he elaborates with a further conclusion, that it is "not possible to follow a rule 'privately'", and his ground here is that otherwise "believing one was following the rule would be the same thing as following it". (PI §202). This adds something further, opening up room for (mis)understanding of what one is doing, for (failure of) mastery of one's *Praxis*.

PI §202 *protects* the I's creative contributions to a practice, rather than adducing any need for consensus of opinions. *My* techniques for employing words may depart from or enrich a practice's shape, but I am *revealed* in the techniques I master and apply (or fail to) in my embeddings of words in forms of life, in my understandings and misunderstandings. I may refuse to adopt the extensionalist point of view, and simply regard the diagonal argument as a method for constructing new decimal expansions: such would not be irrational or without purpose.[28] The *Tractatus*'s remarks on the "I" already suggested that I reveal who I am through the uttering of propositions, comparing them with reality, and acknowledging where this is not done; but for the mature Wittgenstein the variety of moves in language is more complex. The detail of the human organism's *Technik* is left to the side in the *Tractatus*, whereas the mature Wittgenstein's snapshot investigations of language-games draw them out. Language-games allow us to disassemble the workings of language and reflect on *our* masteries and failures of mastery, and only in this way may the constant "convenings" of techniques in use be seen for what they are: confrontations with technics of culture, that is, with others' and my own practices with words, the only way for me to *become* an I (Cavell 1979/1999, 125; 1988; PI §§232, 262).

And now what does it mean to say that a specific form of human activity "belongs to a practice"? One might have thought it to be obvious that whenever one looks at human beings, this is clear. But it is a crucial part of Wittgenstein's mature "anthropological" turn to stress that using our own concepts and rules to identify *which* (form of) practice is being followed (which steps are *steps*) may face its limits.

One might be tempted to think that a specific practice is to be *explained* as a form of activity that is what it is in virtue of being governed by a system of rules which constitute its norms of correctness, or possibly human ends, goods, or practical reasons.[29] But then it would be the notion of *a system of (human end-oriented) rules or norms of correctness or goodness* that explains our understanding of what (and *which* practice) a practice is, not the other way around. One might then notice that every human action we know of, as well as any practice, is open to elaboration, contestation, and interpretation, and so representable in terms of many different rules and points of view, making it suitable for situation and grammatical articulation in a variety of ways.[30] Practices evolve, they are

[28] Floyd and Mühlhölzer (2020), Chapters 7 and 8; Floyd (2020a).

[29] See, for example, Kant (1992), Rawls (1955/1999), n. 1, Frege (1893/1903/2013), Preface.

[30] Cavell (1969/2002,1979/1999); see Narboux (2022) on "elaboratives".

periodically corrected, and they must inculcate new practitioners and may absorb other practices over time. Moreover, simply behaving in ways representable by particular rules is not enough: Without particular techniques of embedding words in life, we might be left with nothing but urgings and stampings of feet, and it is difficult to regard these as part of our practice, though insofar as we regard the stamping of feet as techniques for instantiating our rules, we may (cf. PI §200). Given these complexities, the very notion of locating an action within a particular practice may seem threatened, as well as the very phenomena of necessity and correctness, the (logical) *must* governing the applications of the practice's rules (or words or concepts) from case to case.

Later Wittgenstein is widely understood to have attacked the assumption that practices (or concepts) can be constituted or fixed by closed systems of rules: in his mature thought, it is precisely the plasticity of our and others' elaborations, understandings, and displays of mastery that provides the friction for rule-following to gain its reality (PI §§107, 130; MS 142, 119–120 (1936)). This lies at the heart of his mature "anthropological" turn. Although "mechanical" or "unthinking" steps with rules are realities in the human world, even these take place against backdrops, evolving responses, language-games, human forms of life—a point on which Turing (1936) built his vivid analysis of logic.[31] We *use* human beings *as* machines or living tools for certain purposes (BrBk §67; PI §157), making of them slaves (in Aristotle's sense).[32]

What is puzzling – above all for Wittgenstein – is that what he says

> … seems to amount to this, that logic belongs to the natural history of humankind. And that is not combinable with the hardness of the logical "must". (RFM VI §49)

Although he repeatedly notes certain "very general facts" of human natural history, he cannot defend or justify our "logical 'musts'" in terms of these.[33] Instead, such notes only show that there are certain very general

[31] Floyd (2017, 2023).

[32] The tool is a lifeless slave and the slave is a living tool, as Aristotle (1984) has it: *Nicomachean Ethics* VIII 11; *Eudemian Ethics* VII 9; compare *Zettel* (hereafter Z) §§528–529.

[33] Shaw (2023).

uncontroversial facts and regularites which have so far "never been doubted" (cf. PI §§25, 415, PPF xii). By contrast,

> … the logical "must" is a component part of the propositions of logic, and these are not propositions of human natural history. If what a proposition of logic said was: Human beings agree with one another in such and such ways (and that would be the form of the natural-historical proposition), then its contradictory [*sein Gegenteil*][34] would say that there is here a *lack* of agreement. Not, that there is an agreement of another kind.
>
> The agreement of humans that is a presupposition of logic is not an agreement in *opinions*, much less in opinions on questions of logic. (RFM V §49; cf. RFM VI §30, PI §241)

Even in the *Tractatus* Wittgenstein had denied that one could have opinions about logic: The sentences of logic are tautologous, limiting cases of propositions, and "laws" justifying them would be "superfluous" (TLP 4.462, 5.132).[35] However, in RFM I he elaborates the point in the context of an "anthropological" response to Frege's discussion of logical laws in the Preface to the *Grundgesetze* (hereafter Gg, vi–vii). This response has been widely discussed;[36] I have space only briefly to touch upon large issues here.

Frege imagined a situation arising in which another human or other group—call them logical aliens, following Ricketts (1986)—draw inferences from a set of laws of thought [*Denkgesetze*] that directly contradict [*widersprächen*] ours, so that their judgements about what is true and what follows immediately contradict our inferences and lead to results that are "often" "opposed" [*entgegengesetzten*] to ours. This situation, Frege supposes, cannot be explained by their having made different empirical generalizations in accordance with our logical laws.

A psychologistic or naturalistic account of logic as "the laws of thought" would say that the situation imagined simply contrasts how "they"

[34] Anscombe translates Wittgenstein's "Gegenteil"—"opposite"—as "contradictory", attributing a negation-relation ("Human beings do/do not agree with one another in such-and-such ways"); this use of "Gegenteil" as negation has a basis in Frege 1879, §11; 1918–19/1984, Gg §6, 1897/1997, 149; Shieh 2021 discusses this in relation to the *Tractatus*'s differences with Frege.

[35] Dreben and Floyd (1991).

[36] Ricketts (1986), Cerbone (2000), Nunez (2019), Miguens ed. (2020) and Shieh (2021).

judge vs. how "we" judge: different logical laws for different systems of judgements, analogous to differing conceptions of grammar, or different standards of fashionable dress (Gg xvf). Frege calls such an account "a hitherto unknown form of madness" (Gg xvi), because for him logic consists of the most general laws of truth, those which allow questions of truth to arise at all, that is, those which make room for a (prescriptive) distinction between "holding-to-be-true" and "being true" no matter what the discipline. For Frege all the laws that serve to constitute this distinction must hold of anything we can call—in his non-psychological sense—"thought".

Frege cannot justify the idea that there is no contradiction between our group's judgements and those of the aliens on the ground that each of the two groups follow legitimate yet different laws of thought *and* subscribe to his own conception of logic, that is, the idea of how we *ought* to think, the idea of judging as an action aiming at judgements true or false, judgements capable of contradicting and not contradicting one another, judgements obeying the most general laws of thought (1897/1997, 147). To do so would reduce the idea of laws of logic to merely psychological or group regularities, whereas the laws of logic "mention neither humans nor time" (Gg xvii). The psychologistic logicians, meanwhile, are in the uncomfortable position of having to decide in what way they may or may not say that our logical laws are incorrect or unnecessary or optional or apply to them. But, Frege insists, the issue of with what *right* we recognize logical laws may only be spelled out in logic, not fobbed off elsewhere (to God, transcendental reflection, community agreement, etc.). Frege may reason from the supposition that there are beings who reject, for example, the law of identity, but he confesses that he can neither suppose nor doubt that these beings are *right*: "at least this holds of me" (Gg xvii).

In what way is Frege driven back to what holds "at least of me"—something that he himself would not regard as "logical" at all?[37] Wittgenstein will remark, as we know, that "Frege never told us what this madness would be like" (RFM I §152). He invites Frege to pursue an "anthropological" investigation of *himself*.

[37] Shieh (2021, 22).

He does agree with Frege to a certain extent. In the above-quoted RFM VI §49 he suggests that one ought not make sense of this situation by construing logic as a matter of culture, psychology, or evolved natural history or—one might add—"practice" in any empirical sense. Yet, he suggests, an absence or "lack" of agreement between two different communities' results and ways of judging (in the empirical, natural historical sense) might somehow be overcome *via* a different *kind* of "agreement", one that allows for alienness to rear its head, while at the same time retaining Frege's conception of logic as expressing general forms of thought. To this end, Wittgenstein offers many language-games exploring what the "madness" might be like, elaborating a *variety* of possible techniques of calculating and inferring.

If we scrutinize and compare such differing techniques for embedding words in life, we might revise some of *our* original judgements, for example, "it is nonsensical to give directions on the street to someone by telling them that it doesn't matter which direction they take, left or right, straight or behind, they will get there by any route" (RFM III §§78–79). Humans *can* (≠ will) work at *making* sense of one another: sometimes one needs to invent a technique for embedding a phrase—or for that matter a logical law—in life.[38] Wittgenstein's point is that logic cannot take care of itself without *us*: we can never ultimately avoid the need for mastery, even at home, in applying what we regard as laws of thought, and these are best thought of as conglomerations of techniques with wide purview. The techniques and methods of calculation we use (e.g. counting like *this*) are not mere "pastimes": they are widely used in daily life (RFM I §4) and organically interconnected in a myriad and in general unforeseeable ways. And so they are exposed to different ways of elaborating their parts. Techniques bring about spaces of logical "form" that decompose judging organically, in terms of a variety of modes of proceeding. This "seems to abolish logic" but is in fact what allows for its "forms" or possible applications, which evolve open-endedly, to be maintained (PI §242). The possibility of changes in how we look at and analyze practices

[38] It is perhaps no accident that for Aristotle *technê*, as productive knowledge, deals with particular cases, yet is not simply a knowing how or *Praxis*, but also a knowing how to place and describe a particular case sufficiently to say how it is to be treated (Johansen, ed. 2021, Introduction).

is in fact required in certain cases *for* agreement, *for* logic—classical or otherwise—to apply.[39]

Wittgenstein's responses to Frege on logical aliens in RFM are carefully angled. He does not deny that madness of the kind that concerns Frege might well ensue. But, like Frege, he also does not take every alien confrontation over logic (or calculating) to be a confrontation consisting in a clash of *opinions* that must be resolved, feelings or psychological states, or of separate *practices*. For him, the kinds of "agreements" that logic requires are not always this, not only a matter of "conviction" [*Überzeugungen*] (RFM VI §30). They involve at times "agreement of a different kind", perhaps requiring a change in our understanding of what is being said by whom, or a change of our way of looking at things, or the techniques with which we work. That such changes may lead us astray from truth is an ever-present possibility, but we cannot say this possibility is always realized, given how we do proceed and take ourselves to speak.

For Wittgenstein, Frege left out the importance of *techniques* of thinking (plural), the very idea of different ways of embedding words and symbols in evolving forms of life. Taking this to be fundamental to logic itself allows for the sorts of "madness" that Frege contemplated to be redescribed in an anthropological vein (without anthropological theses). Wittgenstein wishes to leave open the possibility of an evolution in everyone's forms of agreement and judging.

The "homebaked" notion of *techniques*, involved in adjusting means to ends and stitching modes of using language together in criss-cross ways, works against a stipulative "radical" conventionalist reading of Wittgenstein's rule-following remarks according to which we make up anew on each occasion each step that "follows the rule".[40] It also works against the idea that grammatical systems are incommensurable, if only because, as Wittgenstein explicitly remarks (echoing T.S. Eliot and citing F.R. Leavis), it is impossible to say where a technique begins and where it

---

[39] MS 127, 57–60 (1943–44), responding to the open-endedness of Aristotelian forms and mathematical forms of logic; see Floyd and Mühlhölzer (2020, 93–94).

[40] For "radical" conventionalist readings, the *locus classici* are Dummett (1959) and Wright (1980), see also Berg (2024); for criticisms see Diamond (1968/1991, 2004), Stroud (1965), Cavell (1979/1999) and Marion (1998).

ends.[41] It is impossible to say what the *point* of a move is apart from human interests and points to *be* made. *Technique* gives us something to be distinguished from the bruteness of particular expressions of "normative oughts", stipulations, convictions, or regularities that isolate rule-following in particular individual acts, claims, or abstract rails.

Without respecting the work techniques do for Wittgenstein, his later remarks on rule-following and the foundations of mathematics have been interpreted in terms of such brute "normativities", even in very sophisticated anti-conventionalist interpretations.[42] By contrast, the notion of *Technik*, cold though it may sound, opens up space for less bruteness in calculation, more organic forms of criticism and transpositions of thought-movements. The idea is to *link* our own practices with other possible ones *via* investigation of techniques of embedding words in forms of life.[43]

Techniques of Thinking

A crucial remark in RFM further clarifying Wittgenstein's relation to Frege is this:

> The propositions of logic are "laws of thought", "because they bring out the essence of human thinking" – to put it more correctly: because they bring out, or shew, the essence, the technique, of thinking. They shew what thinking is and also shew kinds of thinking (Watson). (RFM I §133=FF §322)[44]

Three grounds for calling the propositions of logic "laws of thought" are canvassed here. The first acknowledges Frege's idea that the laws of thought show us the "essence of human thinking". The second "more correct" view grounds the conception of logic as the laws of thought in

[41] WCL 108 (1939); Floyd and Mühlhölzer 2020, §8.2, LC 37.

[42] See Stroud (1965), Ginsborg (2022), discussed by Shaw (2023, 57ff).

[43] Engelmann and Floyd (forthcoming).

[44] MS 118, 87r (1937), "Watson" deleted in RFM.

the idea of "the technique of thinking". This alludes to Hilbert's metamathematical idea that his logical formula game

> is carried out according to certain definite rules, in which *the technique of our thinking* [*Technik unseres Denkens*, Hilbert's emphasis] is expressed. These rules form a closed system that can be discovered and definitively stated. The fundamental idea of my proof theory is none other than to describe the activity of our understanding, to make a protocol of the rules according to which our thinking actually proceeds. Thinking, it so happens, parallels speaking and writing: we form statements and place them one behind another. If any totality of observations and phenomena deserves to be made the object of a serious and thorough investigation, it is this one - since, after all, it is part of the task of science to liberate us from arbitrariness, sentiment, and habit and to protect us from the subjectivism that already made itself felt in Kronecker's views and, it seems to me, finds its culmination in intuitionism. (Hilbert 1927, 475)

Hilbert here urges the philosophical hegemony of "metamathematics", claiming that only *this* technology allows for comparisons and contrasts to be "objectively" drawn between differing "contentual" views (e.g., classical vs. intuitionistic theories). It will, he promises, use science to "liberate" us from "arbitrariness, sentiment and habit".

We have already seen Wittgenstein resisting the idea that metamathematics can resolve fundamental philosophical questions. In RFM I he corrects Hilbert in giving a third ground for what it is for logic to express "the essence of human thinking": he invokes the idea of *kinds* of thinking, a variety of techniques, reminding himself of Watson, with whom he and Turing had discussed foundations of mathematics and logic in summer 1937.[45]

It now belongs to the "essence" of human thinking, to the essence of logic itself, that there are multiple techniques of thinking. Once again, this idea lies at the heart of Wittgenstein's remarks on the "foundations" of mathematics, his hesitations about cultish remarks about Cantor's diagonal proof. The extensionalist (Dedekind, Cantor, Hilbert, set theorist etc.) *erases* all our humdrum familiar techniques of working with

[45] Floyd (2001, 282ff., 2017).

individual real numbers and rules of calculation so as to incorporate actually infinite objects into mathematics: to complete the continuum, she must. But "puffing this up" (RFM II §21), she would pretend to have transcended the humdrum from whence she came, acting cultishly. Hardy pretends to be "prudish" and scrupulous, while he fails sufficiently to emphasize the tensions involved in the transitions (1929, 1941).[46] Similarly, the claim that *Principia Mathematica* provides a pure *foundation* for arithmetic in logic is cultish insofar as it ignores the fact that we use *multiple* techniques—including ordinary arithmetical techniques—in coming to understand elaborate formal proofs in *Principia* (RFM III, *passim*). There is no "pure" language or *Technik* for logic: the "primitive" techniques or arithmetic express logic as well, and without them *Principia* would be of no interest.

A remark in *The Big Typescript* also illustrates Wittgenstein's path forward:

> (Sraffa) An engineer is building a bridge. To do this, he checks in several handbooks; in technical handbooks and in legal ones. From the one he finds out that the bridge would collapse if he were to make this part weaker than etc., etc.; from the other that he would be locked up if he built it in such and such a way. – Now are the two books not on the same level? – That depends on what kind of a role they play in his life… After all, for him the legal handbook can simply be a book about the natural history of the people around him. Maybe he also has to check in a book on the life of beavers in order to find out how he has to paint the bridge so that beavers won't gnaw at it. – But isn't there still another way of looking at laws? …
>
> The words "the contract binds me" are, to be sure, a pictorial representation, and are therefore a false proposition, given the usual meaning of the word "to bind"; but understood correctly, they are true (or can be true), and they distinguish the one case from another in which a contract merely tells me what it is useful for me to do. And even if one objects to the words "the contract (or the law) binds me", one can't object to the words: "I *feel* bound by the contract". (BT 242)

[46] Floyd and Mühlhölzer (2020, §§5.7,7) and Floyd (2020a).

There are differing ways of looking at what it is to be bound to necessities, and many of these are, strictly speaking, metaphorical. The middle Wittgenstein refuses to agree with Sraffa that the role an expression "plays in one's life" may be understood on "one level" naturalistically, that is, instrumentally, anthropologically, or ethologically.[47] Whereas the more mature Wittgenstein concedes Sraffa's point, while retaining the idea that the contrast between a symbolical expression and an instrumental one marks a difference: it is the multifarious techniques by means of which I express and explain my adherence to the law, making it intelligible in particular cases, that allow for the law's dignity and rigor (RFM I §118).

The manifold character of techniques within every *Praxis* must be respected, and not reduced to something "metalinguistic" about proposed or described uses. When we explain the uses of a word, we are adding to its potentialities: "we describe the technique for using a word in words", and there is always, within any such explanation, "a technique for using an explanation" and for any description "a technique" of applying it (MS 160, 17r–18r (1938)). There is no regress of techniques for techniques, though, because, as *Technik*, they may be repurposed. A novel use may not lead us to accept something as a viable technique, or an appropriate move within *this* practice, but it may be developed as a technique all the same—recall the wayward pupil with 2,4,6.... There is no need in every case to attempt to fashion a word or formula "technology" that will express every step in the practice, and even if we have it, we have not somehow transcended our own variety of techniques and ways of going on.

Acknowledgements I owe a debt to the insightful comments of audiences for early versions of this chapter given at the University of Bergen, University of Paris 1 Panthéon-Sorbonne, Oxford University, University of Pittsburgh, and the University of East Anglia Wittgenstein Workshop in 2022. Special thanks to Thomas Ricketts for posing the sonorous question *Warum Darum*? about the "therefore" in PI (§§201–202) and for the depth of his work on Frege and Wittgenstein, and to Sophia Arbeiter and Marc Gasser-Wingate for their help. Mauro Engelmann and Sanford Shieh have educated me with their writings and

[47] Compare Engelmann (2013a, 172–3).

lively criticisms of several drafts: the support of our ongoing conversations enrich the subject for me whenever we speak; Sanford on Frege is a true beacon. I'm also grateful to Kevin Cahill for his editorial patience.

References

Anscombe, G.E.M. 2000. *Intention*. 2nd ed. Cambridge, MA: Harvard University Press.

Aristotle. 1984. *The Complete Works of Aristotle: The Revised Oxford Translation*. Edited by Jonathan Barnes, Vol. 2. Princeton, NJ: Princeton University Press.

Beaney, Michael, ed. 1997. *The Frege Reader*. Malden, MA: Blackwell Publishers.

Berg, Ásgeir. 2024. Was Wittgenstein a Radical Conventionalist? *Synthese* 203 (2): 37–68.

Bouveresse, Jacques. 1991. 'The Darkness of This Time': Wittgenstein and the Modern World. In *Wittgenstein: Centenary Essays*, ed. A. Phillips Griffiths, 11–40. Cambridge: Cambridge University Press.

———. 2007. Wittgenstein's Critique of Frazer. *Ratio* 20 (4): 357–376.

Cahill, Kevin M. 2011. *The Fate of Wonder: Wittgenstein's Critique of Metaphysics and Modernity*. New York: Columbia University Press.

Cavell, Stanley. 1969/2002. *Must We Mean What We Say? A Book of Essays*, Revised ed. New York: Scribner/Cambridge University Press.

———. 1979/1999. *The Claim of Reason : Wittgenstein, Skepticism, Morality, and Tragedy*. New York: Oxford University Press.

———. 1988. The Argument of the Ordinary: Scenes of Instruction in Wittgenstein and in Kripke. In *Conditions Handsome and Unhandsome: The Constitution of Emersonian Perfectionism*, 64–100. Chicago, IL: University of Chicago Press.

———. 2013. Declining Decline: Wittgenstein as a Philosopher of Culture. In *This New Yet Unapproachable America: Lectures after Emerson after Wittgenstein*, 29–77. Chicago, IL: University of Chicago Press.

Cerbone, David R. 2000. How To Do Things With Wood: Wittgenstein, Frege and the Problem of Illogical Thought. In *The New Wittgenstein*, ed. Alice Crary and Rupert Read, 293–314. New York: Routledge.

Diamond, Cora. 1968/1991. The Face of Necessity. In *The Realistic Spirit*, 243–266. Cambridge, MA: MIT Press.

———. 1999. How Old Are These Bones? Putnam, Wittgenstein and Verification. *Aristotelian Society Supplementary Volume* 73 (1): 99–134.

———. 2004. Criss-Cross Philosophy. In *Wittgenstein At Work: Method in the Philosophical Investigations*, ed. Erich Ammereller and Eugen Fischer, 201–220. New York: Routledge.

Dreben, Burton, and Juliet Floyd. 1991. Tautology: How Not To Use A Word. *Synthese* 87 (1): 23–49.

Dummett, Michael. 1959. Wittgenstein's Philosophy of Mathematics. *The Philosophical Review* 68 (3): 324–348.

Engelmann, Mauro Luiz. 2013a. Wittgenstein's 'Most Fruitful Ideas' and Sraffa. *Philosophical Investigations* 36 (2): 155–178.

———. 2013b. *Wittgenstein's Philosophical Development: Phenomenology, Grammar, Method, and the Anthropological View*. UK, Palgrave Macmillan: Basingstoke.

———. 2016. The faces of 'Necessity', Perspicuous Representation, and the Irreligious 'Cult of the Useful': The spenglerian Background to the Firsrt Set of Remarks on Frazer. In *Wittgenstein's Remarks on Frazer: The Text and the Matter*, ed. Lars Albinus, Josef G.F. Rothhaupt and Aidan Seery, In On Wittgenstein, 129–174. Berlin/Boston: De Gruyter.

———. 2024. Mauthner, the *Tractatus*, and the Kraus Circle. In *44th International Wittgenstein Syposium 2023: 100 Years of Tractatus Logico-Philosophicus—70 years after Wittgenstein's Death, a Critical Assessment*.

Engelmann, Mauro Luiz, and Juliet Floyd. forthcoming. Wittgenstein on Religion as a Form of Life: From a 'Jamesian Type' to Remarks on Frazer. In *Wittgenstein on the Epistemology of Belief*, ed. Nuno Venturinha and Duncan Pritchard. Cambridge: Cambridge University Press.

Floyd, Juliet. 1991. Wittgenstein on 2,2,2…: On the Opening of *Remarks on the Foundations of Mathematics*. *Synthese* 87 (1): 143–180.

———. 1995. Wittgenstein, Gödel and the Trisection of the Angle. In *From Dedekind to Gödel: Essays on the Development of the Foundations of Mathematics*, ed. Jaakko Hintikka, In Synthese Library, 373–426. Dordrecht: Kluwer Academic.

———. 2001. Prose versus Proof: Wittgenstein on Gödel, Tarski and Truth. *Philosophia Mathematica* 3 (9): 280–307.

———. 2016. Chains of Life: Turing, *Lebensform*, and the Emergence of Wittgenstein's Later Style. *Nordic Wittgenstein Review* 5 (2): 7–89.

———. 2017. Turing on 'Common Sense' Cambridge Resonances. In *Philosophical Explorations of the Legacy of Alan Turing—Turing 100*, ed. Juliet and Bokulich Floyd, Alisa, In Boston Studies in the Philosophy of Science, 103–149. Dordrecht: Springer.

———. 2018. *Lebensformen*: Living Logic. In *Language, Form(s) of Life, and Logic: Investigations after Wittgenstein*, ed. Christian Martin, 59–92. Berlin: deGruyter.

———. 2020a. Aspects of the Real Numbers: Putnam, Wittgenstein, and Nonextensionalism. *The Monist* 103: 427–441.

———. 2020b. Wittgenstein on Ethics: Working Through *Lebensformen*. *Philosophy and Social Criticism* 46 (2): 115–130.

———. 2021. *Wittgenstein's Philosophy of Mathematics*. In *in the Philosophy of Mathematics*, ed. Penelope Rush, Stewart Shapiro, and Cambridge Elements. Cambridge: Cambridge University Press.

———. 2023. 'Surveyability' in Hilbert, Wittgenstein and Turing. *Philosophies* 8 (1), 6, Basel: MDPI, online at https://doi.org/10.3390/philosophies8010006

Floyd, Juliet, and Felix Mühlhölzer. 2020. *Wittgenstein's Annotations to Hardy's A Course of Pure Mathematics: An Investigation of Wittgenstein's Non-extensionalist Understanding of the Real Numbers*, Nordic Wittgenstein Studies. Vol. 7. New York: Springer Nature.

Frege, Gottlob. 1879. *Begriffsschrift*: A Formula Language of Pure Thought Modeled on That of Arithmetic. In van Heijenoort, ed. 2000, 1–82.

———. 1897/1997. Logic. In Beaney ed. 1997, 227–250.

———. 1893/1903/2013. *Basic Laws of Arithmetic, Derived Using Concept-Script: Volumes I & II*. Translated by Philip A. Ebert, Marcus Russberg and Crispin Wright. Oxford: Oxford University Press.

———. 1918–19. Negation. In Beaney ed. 1997 346–361. Originally published in *Beiträge zur Philosophie des deutschen Idealismus* I (1918–19), 143–157.

Ginsborg, Hannah. 2022. Going on as One Ought: Kripke and Wittgenstein on the Normativity of Meaning. *Mind & language* 37 (5): 876–892.

Hardy, G.H. 1929. Mathematical Proof. *Mind* 38 (149): 1–25.

Hardy, G. H. 1941. *A Course of Pure Mathematics*. 8th ed., revised. Cambridge: Cambridge University Press.

Heidegger, Martin. 1993. The Question Concerning Technology. In *Basic Writings*, ed. David Farrell Krell, 307–342. New York: Harper Collins. Original edition, 1954, in Heidegger, *Vorträge und Aufsätze*, Stuttgart: Neske Verlag.

van Heijenoort, Jan. 2000. *From Frege to Gödel: A Sourcebook in Mathematical Logic 1879–1931*. Cambridge, MA: Harvard University Press.

Hilbert, David. 1925. On the Infinite. In van Heijenoort, ed. 2000, 161–190. Original edition, 'Über das Unendliche'. *Mathematische Annalen* 95 (1926), 161–190.

———. 1927. The Foundations of Mathematics. In van Heijenoort, ed. 2000, 464–479.

———. 1930. Logic and the Knowledge of Nature. In *From Kant to Hilbert: A Source Book in the Foundations of Mathematics*, ed. William Ewald, vol. II, 1157–1165. New York/Oxford: Oxford University Press.

Johansen, Thomas Kjeller, ed. 2021. *Productive Knowledge in Ancient Philosophy: The Concept of Technê*. Cambridge: Cambridge University Press.

Kant, Immanuel. C. 1770–90/1992. *Lectures on Logic*. Edited by J. Michael Young. *The Cambridge Edition of the Works of Immanuel Kant*, edited by Paul Guyer and Allen W. Wood. Cambridge: Cambridge University Press.

Marion, Mathieu. 1998. *Wittgenstein, Finitism and the Foundations of Mathematics*. New York; Oxford: Clarendon Press.

Miguens, Sofia, ed. 2020. *The Logical Alien: Conant and His Critics*. Cambridge, MA: Harvard University Press.

Musil, Robert. 1921/1978. Mind and Experience: Notes for Readers Who Have Eluded the Decline of the West. In *Precision and Soul: Essays and Addresses*, ed. Burton Pike and David S. Luft, 134–149. University of Chicago Press. Original publication, "Geist und Erfahrung", *Der neue Merkur*, Mar. 1921 8:1042–59.

Narboux, Jean-Philippe. 2022. Actions and Their Elaboration. In *Stanley Cavell's Must We Mean What We Say? at Fifty*, ed. Gregory Chase, Juliet Floyd, and Sandra Laugier, 56–76. Cambridge Philosophical Anniversaries. Cambridge: Cambridge University Press.

Neurath, Otto. 1921/1973. Anti-Spengler. In *Empiricism and Sociology*, ed. Marie Neurath and Robert S. Cohen. Vienna Circle Collection, 158–213. Boston, MA/Dordrecht: D. Reidel. Original edition 1921, *Anti-Spengler*, Munich: Georg D. W. Callwey.

Nunez, Tyke. 2019. Logical Mistakes, Logical Aliens, and the Laws of Kant's Pure General Logic. *Mind* 128 (512): 1149–1180.

Rawls, John. 1955/1999. Two Concepts of Rules. In *John Rawls: Collected Papers*, ed. Samuel Freeman, 20–47. Cambridge, MA: Harvard University Press. Original edition, *Philosophical Review* 64, 1 (January 1955): 3–32.

Ricketts, Thomas G. 1986. Objectivity and Objecthood: Frege's Metaphysics of Judgment. In *Frege Synthesized*, ed. Leila Haaparanta and Jaakko Hintikka, 65–95. Dordrecht: Springer Netherlands.

Schatzberg, Eric. 2018. *Technology: Critical History of a Concept.* Chicago: University of Chicago Press.

Shaw, James R. 2023. *Wittgenstein on Rules: Justification, Grammar and Agreement.* Oxford: Oxford University Press.

Shieh, Sanford. 2021. What Could Be the Great Debt to Frege? or Gottlobius ab paene omni naevo vindicatus. *Disputatio, Philosophical Research Bulletin* 10 (18, Sept.): 5–62, online at https://disputatio.usal.es/vol-10-no-18/.

Soulez, Antonia. 2019. Neurath(A). *L'Encyclopédie Philosophique.* https://doi.org/https://encyclo-philo.fr/item/189.

———. 2023. Otto Neurath: La force d'une pensée constructiviste du possible contre le déclinisme. Manuscript.

Spengler, Oswald. 1923/1926/1928. *The Decline of the West: Volume I, Form and Actuality* (1926) and *Volume 2 Perspectives of World History* (1928), translated by Charles Francis Atkinson, New York: Alfred A. Knopf. Original publication *Der Untergang des Abendlandes: Umrisse einer Morphologie der Weltgeschichte.* Munich: Beck (vol. 1, revised ed. 1923; volume 2, 1922). Vol I of Atkinson translation is online: https://www.gutenberg.org/cache/epub/72344/pg72344-images.html, as are the corresponding German editions at https://www.projekt-gutenberg.org/spengler/unterga1/titlepage.html (vol. 1) and https://www.projekt-gutenberg.org/spengler/unterga2/index.html (vol. 2). References are to volume, chapter section and subsection of the English translation.

———. 1932. *Man and Technics: a Contribution to a Philosophy of Life.* Translated by Charles Francis Atkinson. New York: Alfred A. Knopf from *Der Mensch und die Technik, Beitrag zu Einer Philosophie des Lebens.* Munich: C.H. Beck (1931).

Stroud, Barry. 1965. Wittgenstein and Logical Necessity. *Philosophical Review* 74: 504–518.

Turing, A.M. 1936. On Computable Numbers, with an Application to the *Entscheidungsproblem. Proceedings of the London Mathematical Society* 2 (42): 230–65; Corrections: *Proceedings of the London Mathematical Society* 2, 43 (1937): 544–6.

Wittgenstein, Ludwig. 1921. *Tractatus Logico-Philosophicus.* Translated by C.K. Ogden. International library of Psychology, Philosophy, and Scientific method. London/New York: Routledge & Kegan Paul. First German edition in *Annalen der Naturphilosophie,* Wilhelm Ostwald, ed., 14, 1921, 12–262.

———. 1969. *Preliminary Studies for the 'Philosophical investigations': Generally Known as the Blue and Brown Books*. [BB, BlB, BrB] Edited by Ludwig Wittgenstein, 2nd ed. Oxford: Basil Blackwell.
———. 1970. *Zettel*. [Z] Translated by G. E. M. Anscombe. Edited by G.E.M. Anscombe and G.H. von Wright. Berkeley, CA: University of California Press.
———. 1978. *Remarks on the Foundations of Mathematics*. [RFM] Edited by G. H. von Wright, Rush Rhees and G. E. M. Anscombe. revised, from 1956 edition ed. Cambridge, Mass: M. I. T. Press.
———. 1980. *Philosophical Remarks*. [PR] Chicago: University of Chicago Press.
———. 1989. *Wittgenstein's Lectures on the Foundations of Mathematics: Cambridge, 1939*. [LFM] From the notes R. G. Bosanquet, Norman Malcolm, Rush Rhees, Yorick Smythies, Edited by Cora Diamond. Chicago: University of Chicago Press.
———. 1993a. *Philosophical Occasions, 1912–1951*. [PO] Edited by James Carl Klagge and Alfred Nordmann. Indianapolis: Hacket Publishing.
———. 1993b. "Remarks on Frazer's *Golden Bough*." [RFGB] In Wittgenstein 1993a, 118–155.
———. 1998. *Culture and Value: A Selection from the Posthumous Remains*. [CV] Revised second edition with English translation. Edited by G.H. von Wright in collaboration with Heikki Nyman, revised edition of the text by Alois Pichler, translated by Peter Winch. ed. Oxford, UK; Cambridge, MA: Blackwell.
———. 2000. *Lectures and Conversations on Aesthetics, Psychology and Religious Belief*. [LC] Revised ed. Edited by Cyril Barrett from the notes of Yorick Smythies, Rush Rhees and James Taylor. Berkeley, CA: University of California Press.
———. 2001. *Ludwig Wittgenstein Philosophische Untersuchungen Kritisch-genetische Edition*. Edited by Joachim Schulte, Heikki Nyman, Eike von Savigny and G.H. von Wright. Frankfurt am Main: Suhrkamp.
———. 2003. *Ludwig Wittgenstein: Public and Private Occasions*. [PPO] Edited by James C. Klagge and Alfred Nordmann. Lanham, Md: Rowman & Littlefield Publishers.
———. 2005. *Big Typescript, TS 213*. [BT] Edited by C. Grant C. Grant Luckhardt and Maximilian A.E. Aue. Malden, MA: Blackwell Publishing.
———. 2009. *Philosophische Untersuchungen = Philosophical investigations*. [PI] Rev. 4th ed. Edited and Translated by G. E. M. Anscombe, P. M. S. Hacker,

and Joachim Schulte. Chichester, West Sussex, U.K.; Malden, MA: Wiley-Blackwell.
———. 2014. *Wittgenstein: Lectures, Cambridge 1930–1933, From the Notes of G.E. Moore.* [WML] Edited by David Stern, Brian Rogers and Gabriel Citron. Cambridge: Cambridge University Press.
———. 2015. *Wittgensteinsource: the Bergen Nachlass Edition: www.wittgensteinsource.org.* Edited by Alois Pichler. Bergen: Wittgenstein Archives, University of Bergen.
———. 2017. *Wittgenstein's Whewell's Court Lectures, Cambridge, 1938–1941: From the Notes of Yorick Smythies.* Edited by Volker A. Munz and Bernhard Ritter. Oxford: Wiley-Blackwell.
———. 2020. *Ludwig Wittgenstein: Dictating Philosophy to Francis Skinner—The Wittgenstein-Skinner Manuscripts.* In *Nordic Wittgenstein Studies*, ed. Arthur Gibson and Niamh O'Mahony. Springer Nature: Switzerland.
Wright, Crispin. 1980. *Wittgenstein on the Foundations of Mathematics.* London: Duckworth.
von Wright, G.H. 1983. Wittgenstein in Relation to His Times. In *Wittgenstein*, 201–216. Minneapolis: University of Minnesota Press.

Mathematics and Aspect-Seeing

Reshef Agam-Segal

How does Wittgenstein's late discussion on aspects bear on his philosophy of mathematics? Can the former shed light on the latter, suggest refinements?[1]—In the *Investigations*, stressing the philosophy of mathematics, Wittgenstein says that philosophy should leave everything as it is,

[1] The editors of Wittgenstein's volumes on the philosophy of psychology (*RPP1, RPP2, LW1, LW2*), which contain source material for his discussion about aspect-seeing in the second part of the *Investigations*, and more recently the editors of the 4th edition of the *Investigations*, who retitled the second part, and called it "Philosophy of Psychology—a Fragment," have created the impression that the relevancy of the discussion about aspects is more or less limited to the philosophy of psychology. This is misleading (see Diamond 1984; Knott 2017, 2020; also Rush Rhees' preface to BB). My disagreement is not merely historical, regarding Wittgenstein's intentions; it is primarily philosophical: The proof that the relevancy of the discussion of aspects goes well beyond the philosophy of psychology is in the pudding of demonstrating how understanding the grammar of aspect-seeing helps clarifying issues in other philosophical areas. Beside the philosophy of mathematics (see also Floyd 2010; Zwicky 2010), the phenomena of aspects are relevant to aesthetics (Wollheim 1980), ethics (Agam-Segal 2018), philosophy of perception (Travis 2016), philosophy of language (Zwicky 2014), and metaphilosophy (Agam-Segal 2012a, 2015). Wittgenstein himself wrote explicitly about aspect-seeing and related issues in connection to all those philosophical areas.

R. Agam-Segal (✉)
Virginia Military Institute, Lexington, VA, USA

K. M. Cahill (ed.), *Wittgenstein on Practice*,
https://doi.org/10.1007/978-3-031-68656-6_9

be descriptive (*PI*, §124). But description is not innocuous; redescription can change our whole perception and conception of an issue—to gesture at the phenomena of aspect. Philosophy generally can be seen in this light, that is, as an activity of redescriptions: What we *need* philosophically, the balm to the philosophically aching mind, is a changed look at things. Not only in philosophy: "In Mathematics, Ethics, Aesthetics, Philosophy, answer to a puzzle is to make a synopsis possible" (*LC*, 358; see also *PO*, 181; *CV*, 55). My focus here will be on mathematics; for mathematics too involves redescriptions. Specifically, I'll argue, some proofs—those that we can follow more than merely algorithmically—reveal aspects; they generate mathematical insight, not just knowledge, as they have the power to make us see things anew.[2]

The second section below explains what aspect-changes are and how some mathematical proofs provoke them, or something very similar. This requires some background, in the first section, about how mathematics may shape the way we see things and thus prepare practices. I distinguish between the way mathematical propositions are preparatory and the way proofs are. The third section rejects the view that aspects are relevant to mathematics only insofar as picture-proofs are involved; algebraic-proofs too, or so I argue, may have the power to reorient our conceptions. The fourth section argues that changing an aspect in mathematics involves an experience of our prelinguistic, pre-mathematical, "primitive," sense of things. This rudimentary sense, contingent as it is, guides the formation of mathematical practices and concepts and grounds them. The fifth section concludes by positioning the view between realism and antirealism. Throughout, I use simple mathematical examples, so that the discussion doesn't get distracted by technical difficulties.

[2] Wittgenstein explicitly compares philosophy and mathematics in *PI*, §144: "I wanted to put that picture before him, and his *acceptance* of the picture consists in his now being inclined to regard a given case differently: that is, to compare it with *this* rather than *that* set of pictures. I have changed his *way of looking at things*. (Indian mathematicians: 'Look at this.')" For further comparisons between philosophy and mathematics see Floyd (2000), and footnote 19 below and the text leading to it.

Mathematics as Preparatory

Mathematics can generate a sense of dazzlement and disorientation. We sense it for instance when we first go through Cantor's diagonal proof, or when we first learn about imaginary numbers, and even of negative numbers if we can recall the experience. It is tempting to construe this disorientation as a symptom of ignorance. Thus, Cantor takes himself to have discovered a fact yet unknown: "I have discovered the difference between a so-called continuum [of all real numbers which are ≥0 and ≤1] and any set like the totality of real algebraic numbers" (1996, 841). But, I'll argue, this experience is better understood as the experience of our concepts being molded, or recast.[3] With that, our orientation to things shifts, our ontological sense of them transforms. Our mind's very interface with the world, so to speak, is thus being shaped or reshaped in mathematics. The disorientation is a result of reorientation, so to speak, and mathematics is thus preparatory: it epistemically (re)positions us for thinking and dealing with things—quantities, shapes, lengths, etc. "Mathematical and logical propositions are still *preparations* for a use of language" (Wittgenstein *LFM*, 249).[4] Let us distinguish between the way mathematical propositions are preparatory, and the way proofs are.

Mathematical Propositions as Preparatory

Consider '1 + 1 = 2': Suppose we take an apple, then another, and count them. As apples do, neither vanishes nor multiplies, and we all, or most of us, reach the same result: two. Doing it with enough apples, we get an empirical generalization: 'If you add one apple to another, you get two.' But now, the same happens with books, tables, cats, and many other

[3] Cantor's proof, for instance, can be understood not so much as relieving our ignorance about the number of real numbers relative to natural numbers, which is one standard understanding of the proof—as if it is already clear what it even means for the reals or the naturals to have a number, a quantity. Rather, among other things, the proof demonstrates certain things about the very concepts *number, quantity*: for one thing, how these concepts are not usefully thought of as the same in different contexts; how they are family-resemblance concepts.

[4] See also Wittgenstein *RFM*, I, §72, where he talks of a formula to statements that go on to make use of a word. There are also non-preparatory conceptual investigations; these are more common in aesthetics, for instance; see Agam-Segal (2012b).

things. True—and this is important—it doesn't happen with all things, or not always: not with clouds, raindrops, sensations, shadows, curls in people's hair, sunbeams, wrinkles on elephants, and flickering tongues of flame in campfires; these seem to have different ontologies, which our mathematically regimented number concepts don't quite capture. Still, what happens with apples happens with many other things, which allows us to view our empirical generalization as representative. An abstract equation forms: '1 + 1 = 2,' but now not as a further empirical generalization. It is "proven" (informally) by what happens with the apples and the other things in the following sense: it faithfully captures the quantitative ontology of apples, books, tables, cats, also hours, vice-presidents, symphonies, etc. The kind of ontology these things have is mirrored and expressed by '1 + 1 = 2'; the proposition "characterizes" what these things are (see Floyd 2018). The ontology of (very many) things is thus studied via a mathematical study, and mathematics is an ontological-conceptual study, studying things and their ontological features—quantities, relations, shapes, etc.—by studying their proper mathematical conceptualization, mathematization. Epistemically, 1 + 1 = 2 captures our sense of that ontology. It is a *face* of very many things: a Gestalt, an abstract image through which we may view quantifiable phenomena—an ontological essence of them. It allows us to 'see into the possibilities of things,' set our expectations, capture things in thought, position them intellectually, be at home with them, and have them in our world. It is a readymade way for talking about things, a form of representation, and a useful rule to have.[5]

Now, we often don't pause to pay direct attention to such Gestalts; we typically rather take them for granted. That things have such and such a face (ontology) for us comes out in how we handle them—for example, the fact that we count apples without worrying that they might merge or vanish. We are not astonished when they don't. This lack of

[5] In a related context, Wittgenstein suggests that mathematical propositions can be thought of as hardened experiential propositions (*RFM*, IV, §§22–23). When a proposition hardens, as I understand it, it changes its function—it becomes a rule. Apart from what I mention in the text, this also involves people practicing in the same way—a convention forming. It might also be mentioned that as part of the proposition losing its experiential status and becoming abstract and timeless, the result of such hardening is no longer taken to capture an external relation: The relation between the two sides of the equation is now rather taken to be internal, conceptual (Wittgenstein *RFM*, III, § 55). See also Bangu (2019).

astonishment, this taking it for granted, shows how mathematical truths capture our ontological sense of things, and animate our regard to them even from behind the scenes, so to speak. To wit, like the ontological sense they capture, the mathematical truths that capture it are often not themselves objects of direct awareness. They are like flashlights that allow us to see things but are themselves not in view (to borrow a metaphor from Simone Weil, 1970, 147). Nevertheless—and below I connect this to aspect-seeing—Gestalts can be revealed; they can dawn, come on stage. And when that happens, both *what* things are and *how* we organize them in thought become an issue. Pointing out that 23% of 50 equals 50% of 23, for instance, reveals the (different) faces quantities may have, makes us pay attention. In such ways, Gestalts are preparatory.

Proofs as Preparatory

Mathematical propositions reveal aspects of the ontology of (very many) things, the quantitative life of apples and books for instance, and how such things make quantitative sense to us. But sometimes we want to examine how they *should* make sense: whether our theorems accurately capture things' ontological life. This is what proofs are for. Now, sometimes, all this means is that the proof shows algorithmically how a theorem is consistent with the rest of mathematics, thus $x^0 = x^{1-1} = x^1 \times x^{-1} = x^1/x^1 = 1$: Exponentiation typically means 'repeated multiplication'; but what is this supposed to mean when the power is 0? The proof guides us, tells us what algorithm to follow. This takes us so far; it gives us knowledge what to do in certain cases, helps us avoid problems, and perhaps in this case that's all we need. In other cases, however, the algorithm does more: it brings to light aspects of the ontology of things—quantities, shapes, operations, relations, etc. Cartesian analytic geometry, for instance, can sharpen our view not only of the mathematics but also of the ontology of shapes. The Jordan curve theorem, to give a more specific example, captures mathematically (part of) the ontology of the relations 'inside' and 'outside.' It says that every continuous path connecting a point inside a closed curve, such as a circle, to a point outside it intersects with the curve, and thus that closed curves divide the plane into two regions: exterior and interior. The proof algorithmizes a pre-mathematical intuition, but also sharpens pre-mathematical concepts: 'inside' and 'outside.' It thereby refines our ontological sense of these relations and, by extension, of things that may stand in such relations. Whether the proof merely offers an algorithm or does more to sharpen our mathematical insight into the ontology of things—quantities, shapes, relations, etc.—can be a matter of degree, and largely depends on us: our familiarity with the relevant mathematical concepts and their applications, and our interest and ability to go beyond merely following the algorithm and rethink the ontology of things.

De Villiers (1990) lists several roles that proofs have; one of them is to verify that our theorems are true. Many think that we cannot be sure they are true without proofs (see De Villiers 1990, 18). But is that so? And if we deny this, are we necessarily saying that proofs only give us psychological satisfaction and are not strictly speaking needed? Following the above, I want to reject both alternatives: We may sometimes know that our theorems are true without formal proofs—as '1 + 1 = 2' can be said to be informally "proven" by what happens with apples and tables when we add them. Such "proofs" disclose the theorem as adequately capturing our sense of things' ontology, and as useful; its truth is thereby *shown*. At the same time, formal proofs are not merely there for psychological satisfaction. Sometimes we don't know whether, or see how, propositions can guide thought, or whether they should—we want to (1) verify that our theorems algorithmically fit within broader schemes of concepts and operations, and perhaps also that (2) they properly reflect our ontological sense of things. Proofs can help with that; they help us into a sort of reflective position from which to examine, establish, and possibly revise, mathematical practices.[6]

[6] Wittgenstein says that proofs are surveyable, or perspicuous (*RFM*, I, §154, III §1). My claim is not about how proofs themselves are seen, but about how proofs allow us to see other things: "Can't we say: the figure which shews you the solution removes a blindness, or again changes your geometry? It as it were shews you a new dimension of space. (As if a fly were shewn the way out of the fly-bottle)" (*RFM*, I, §44; also *LFM*, 53-5). The matters—the surveyability of the proofs themselves and how they remove blindness—are connected: The proof does not just establish the rules for using a certain sign, for example, but discloses the *point* of so using it—reveals a life of the sign; and it does that only insofar as the point of the transitions in the proof are themselves clear. Much of the discussion below is about how proofs allow us to come to see the point, the life, of practices. For a discussion about the surveyability of proofs see Mühlhölzer (2005).

Saying that proofs prepare practices suggests that they are not movements *in* practices, but a kind of meta-activity to establish practices: not an "employment of a sense," but "determination of a sense" (*RFM*, III, §37). Proofs sometimes have an impact on the very meanings of the expressions they employ: we cannot assume we fully know already, before the relevant proofs, what 'trisection' or 'numeric-size' mean in the cases of 'trisection of angles' and the size of infinite sets, or what 'inside' means before we prove the Jordan curve theorem: Is the tea strictly speaking 'inside' the cup, even though it is not surrounded by the cup on all sides—not from above? Maybe, strictly speaking, we should say that it is 'on' the cup, as the cake is on the plate? What exactly is the ontology of inside and outside?—Now, this raises a worry: We very much feel we know the meanings of the expressions that feature in our proofs; we don't choose them at random. Besides, if those meanings are not already fixed, then there is seemingly nothing to guide our proofs and we can infer more or less as we like. Such "anarchism" would undermine the very possibility of mathematical normativity. Michael Dummett found Wittgenstein's views to be so anarchist (1978, 64), and objected: "[F]ace to face with a proof, we have no alternative but to accept the proof if we are to remain faithful to the understanding we already had of the expressions contained in it" (1959, 333). Let me examine the matter with a relatively simple test-case: the impossibility of dividing by zero.

Preliminary note: Things go sideways in different ways when we try to divide by 0. The following only explores one such way. Others involve trying to determine uniquely the value of 0/0, or introducing a new sign, say '∞,' and trying to define the rules for it. Generally, different proofs of the same theorem explore different aspects of an issue—different Gestalts (or lack thereof in the case of impossibility proofs, like the present one) and implications.

So why can't we divide by zero? We may view the matter as a problem in which there is *something* we cannot do. So far, our wish to know why sounds reasonable. But how good is our grasp of the question? Let's face it, we don't have a stable enough idea yet of what 'dividing by zero' even means, what *activity* we want to perform and can't; we have an expression in our question whose meaning isn't fully clear. So we first need to ask what 'dividing by zero' means; what activity it designates. Hopefully, once we know that, we'll be able to go back to our question.

To restart, then, let's try and do it: divide by zero; maybe this will show what it means. Here is one way of attempting that: Take '$a \times 0 = b \times 0$'; this is true for any *a* and *b*. Now the rule says that if we have the same multiplier on both sides of an equation, we can divide by it: drop it from both sides. We now get $a = b$, and this, to recall, for *any a* and *b*, which, it is plain to see, means chaos. If $a = b$ for any *a* and any *b*, then nothing makes numerical sense any more, nothing has recognizable quantitative life. Question: Did we really divide by zero? We may want to say 'yes': we followed the rules for division. True, we did not cut anything or distribute a quantity—we did not perform any of *these* activities of dividing, but we nevertheless did something that is akin to one thing we sometimes do when we divide: we dropped a multiplier on both sides of an equation. But, on the other hand, note what happened: Following our existing rules for dropping a multiplier, we connected the expression $x/0$ to what we already take to be meaningless, namely '$a = b$ for *any a* and *b*.' Amid following the rules of division, we thus lost the sense of dividing—the sense that we are doing anything really; an otherwise meaningful activity disintegrated into a mere lifeless following of an algorithm, mere movements without sense. Thus, there is a serious sense in which dropping that multiplier and erasing those zeros, which if we wish we may still call 'dividing by 0,' is shown not to amount to anything clear; no *mathematical* activity was involved, no *dividing*.

We can now look back at what we did and see it again. At first, we thought we were doing something when we erased the zeros; we even gave it a name, 'dividing by zero.' But it now looks different. The situation is like "the fairy tale gift that seems magical in the enchanted castle and if it is looked at in daylight is nothing but an ordinary bit of iron" (*CV*, 11). Dividing by zero revealed itself to be a mere shell of an action, a mere erasing of some numbers from a page; "It is a profitless performance!" (*RFM*, I, appendix 3, §12, see also *PI*, §426). Importantly, we *can* do it, but merely 'algorithmically' so to speak; we cannot really mean anything mathematical by it. There is nothing there for us to mean. Such is the ontology of dividing. 'Dividing by zero' is 'impossible' in the same sense that it is impossible to remember the future, eat backwards, sit quickly, or draw a diameterless circle: these expressions simply mean nothing (which doesn't imply we cannot find a use for them: Ed Dain told me that his son used to talk of 'dividing by zero' as if it meant not dividing at all—making zero divides: so 218 divided by zero would be 218. Dividing by one then meant dividing once: so splitting in half. Arguably, similar things happened when we discovered a way of talking of 'subtracting 3 from 2,' or of 'the square root of -1'). It anyway seems that we don't need to go back to our original question after all; ontologically, it has nothing to be about. Wittgenstein generalized:

> For many mathematical proofs do lead us to say that we *cannot* imagine something which we believed we could imagine. (E.g., the construction of the heptagon.) They lead us to revise what counts as the domain of the imaginable. (*PI*, §517)

Like other impossibility proofs—concerning heptagon-construction, angle-trisection, or circle-squaring, etc.—our proof thus *negatively* prepares the practice, namely withdraws an expression, *x*/0, from circulation (*PI*, §500), which reflects an ontological realization: there is nothing for the expression to correspond to, to mean. Positively, the expression 'dividing by zero' becomes a sort of linguistic signpost, saying "don't go there"; it functions as a limit. And I want to again emphasize that this signposting captures our ontological sense of division, the life that the practice has for us, and it does so by identifying where the practice becomes lifeless. This, as we can reveal the point of the practice of listening—the ontology of 'listening'—by noting that listening to the sound of red is meaningless; listening becomes lifeless there. The mathematical proof here, then, does not just establish the rules, the algorithm, of division, but also discloses the practice's point—reveals the ontological sense that dividing has for us. In Dummett's terms, the proof shows what remaining faithful to our terms demands—that is, how to faithfully introduce further determinacy into our practices.[7]

---

[7] Parallel claims can be made for other impossibility proofs, for example, about trisecting angles. There, we can translate the expression 'Euclidean trisections of an arbitrary angle' into algebraic terms using Cartesian analytic geometry, and ultimately connect it to meaningless expressions, for

Did we not, however, follow existing rules in the course of the proof? We did; but ultimately the rules gave in; they were unable to make mathematical sense of the obscure expression we had in our question. This, I argued, faithfully reflects our sense that no ontology corresponds to $x/0$—no quantity, no division-activity. And, like a Heideggerian hammer that breaks amidst hammering, we naturally stopped and looked at the rules—the division-rules through the lens of which we previously made sense of things and handled them, the rules that until that point did a good job capturing the ontology of things. These rules, which used to be among the conditions for seeing, thinking, and sense-making, now became themselves objects of attention and reflection. In this way, the proof propelled us to a reflective, preparatory, position. Here, as I explain below, we were guided by natural predeterminate, preconceptual, pre-mathematical sensitivities—by our rudimentary ontological sense of things.[8]

To emphasize: the preparation is made possible by the proof. We did not get to that reflective, pre-mathematical, preparatory position just by asking a meta-question about the rules. Rather, the proof took us to that position, and showed us what we can see from there—put us face to face with our rudimentary ontological sense of things. Before the proof, we may have had some vague idea of the possibility of a position from which to examine our rules; the proof gives us a clear notion of that position, and puts us in it.

Aspect-Seeing and Proofs

Mathematical proofs, I argued, put us in a reflective, preparatory, position, and I want to look further at the nature of this position. It is, I'll argue, the same kind of reflective position we occupy when we see aspects. So let me explain what aspect-seeing involves.

---

example, 'the rational roots of $8x^3 - 6x - 1 = 0$.' For more on Euclidean trisections, and other impossibility proofs, including Gödel's theorems, see Floyd (2000).

[8] There is no need to think of this as involving awareness of a *Ding an sich*. Nevertheless, there is indeed a nonconceptualist element in my proposal: there are different kinds of preconceptual awareness. Below, for example, I discuss how "primitively" grouping together different shades of blue allows the concept 'blue' to form. For more on the debate about conceptualism in a related context, see Baz (2019), and Agam-Segal (2023).

Seeing aspects involves a reflective, possibly reorienting, experience of the manner in which things are situated for us, possibly conceptualized. Seeing an aspect is thus more than seeing something that is already routinely and unproblematically situated; it is unlike ordinary seeing, and is not part of it.[9] I follow Avner Baz's use of the term (2000). This opposes a received account in Strawson (1974), Mulhall (1990), and many others, so let me briefly review the opposition. Roughly, the Strawson-Mulhall view is that (continuous) aspect-seeing characterizes our routine seeing of things: Things are normally animated for us when we see them, they point out. Our visual regard to them is typically not algorithmic. Typically, we don't merely *know* what we see without actually *seeing* it; our visual experience does not typically fail to be animated in the relevant way, as happens with picture-puzzles, when we are told what we are supposed to see but can't. We don't typically see a chair, for instance, and wonder what it is. On the Strawson-Mulhall view, this animatedness suggests that aspect-seeing is routinely involved in seeing. My position, following Baz, does not deny this distinction between mere knowing and seeing, or between animated and nonanimated-algorithmic seeing; the Strawson-Mulhall view is right about that. However, I take this issue to be separate from the discussion about aspects. Experiencing aspects, I hold, describes not some routine unnoteworthy seeing of something as already taken for granted, animated as a matter of course, unproblematically embedded in some expected familiar context. The term 'aspect-seeing' rather captures a special phenomenon; a special kind of animatedness. It involves a non-routine experience of something breaking away and becoming animated in a novel way or anew—of a concept for example (but sometimes a preconceptual form of awareness), as it is 'spreading a new ordering veil' over an object to use Wittgenstein's metaphor (*RPP1*, §961).[10] An image

[9] This too, I suggest, is what Wittgenstein means when he says: "'Seeing as....' is not part of perception" (*PI*, 197).

[10] What may confuse is that aspect-seeing can be preparatory, that is, lead to practice-determination, conceptualization, a determination of a routine, which is essential in mathematics. This Baz does not note. It is then natural to say that an aspect got embedded in a routine, and then speak of visual awareness of the aspect in determinate practice, that is, not in experience. Note, however, that such talk has a point only if it alludes to some shift or novelty. For example: "whoever invented calculation in the decimal notation surely made a mathematical discovery. But could he not have made this discovery all in Russellian symbols? He would, so to speak, have discovered a new *aspect*" (*RFM*, III, §46). The Strawson-Mulhall view is not sensitive to this. For discussion, see my 2019.

of being inside and outside language may help: When we ordinarily, routinely, use words like 'duck' or 'salt' of what we see, our position is internal. We see matters through the routines of language, through the concepts. The concepts animate what we see (not merely know). According to my use of the term, no aspect-seeing is then happening. The grammatical routines and concepts involved then function like flashlights that allow us to see, or like cogwheels to which our minds attach and turn. Seeing is often like that. When we identify a duck on the river, or ask for the salt during dinner, we don't wonder about the words 'duck' or 'salt'—how these words work, and whether they suit our needs. We simply use them—make our minds manifest through them. The words are tools—extensions of our mind, so to speak—like that Heideggerian hammer, which extends our arm. Now, like the hammer or flashlight, concepts are routinely tools for seeing, thinking, and sense-making; that is, they remain in the background and are typically not themselves objects of sight or attention. Again, we don't look *at* the flashlight, but *with* it; it makes things seen by remaining itself unseen. But sometimes, and this is typical with aspect-seeing, our concepts come to the foreground and become the object of attention; we suddenly see not through them, but *them*. Seeing the duck in , for instance, seeing that physiognomy, is also attending to how the concept 'duck'—this particular possible judgment—may involve itself with our experience, perhaps even being surprised by the fact that it does (Wittgenstein *PI*, 199); this, importantly, without compelling us to make any actual judgment. We then occupy a sort of reflective position with regard to that concept—think *about* it, not just *with* it. Similar things happen when a weeping willow strikes us as sad, or a melody as plaintive. This too involves not just experiencing something, but also—and as part of the experience—how that thing is reflectively, nonroutinely, mentally positioned for us. Furthermore, this sort of thing doesn't only happen with concepts, as I mentioned before, and isn't necessarily visual. We may similarly be struck when hearing our mother's voice in our own (imagine yelling at your daughter), or recognize a boy we once knew in the face of a middle-aged man; here no concept is involved. In all these cases, our position is reflective, and thinking takes a special form (see my 2014).[11]

The position we occupy in aspect-seeing is also reflective about our ontological sense of things—*what* they are, their quiddity. To suddenly see the duck in  after we took it for a rabbit-picture is, I suggest, also a form of wondering about what it is—a rabbit-picture or a duck-picture. Hearing our mother's voice in our own is also wondering about ourselves—who we have become, what we can't escape.[12] The reflective position we occupy in aspect-perception allows us to examine the way in which we are oriented towards things, and simultaneously how this orientation determines what we have in our world. These are sides of the same coin.

Going back to mathematics, my claim is that the situation here is similar. In mathematics too, we have routines conventions and concepts through which we regularly think about things, and which capture what ontological life they have for us. We count apples, for example, without worrying that they might merge or multiply; we measure clothes without worrying that they might change shape. Such activities reflect a sort of ontological trust on our part: a kind of mental orientation. Now, typically, our mental orientations or attitudes, as reflected in our practical-conceptual routines, guide us: our routines make things visible or thinkable, but are themselves unattended. Sometimes, however, they become objects of attention—for example, when a proof forces us to look—and then we notice and possibly shape our mathematical orientation to things from a reflective point of view. As the duck-experience puts us in a reflective position about , so the proof regarding $x/0$, if followed not merely algorithmically, makes us reflect on what quantities and division are, and indirectly also on the ontology of things that may have a quantity or be divided. It would not be a huge stretch to say that mathematical proofs, when followed more than merely algorithmically, reveal aspects: ways of orienting ourselves towards things.

[11] Diamond (1991, 250) makes a similar distinction between being struck by a new physiognomy, and taking things as a matter of course.

[12] Such experiences involve perceptual, that is, not purely intellectual, wondering. For discussion about the very idea of perceptual reflection see my 2014.

Three comments:

1. The kind of aspect-experience that the proof regarding *x*/0 generates does not concern a specific physical object, as is the case with . Mathematics treats of our modes of representation and thought in a general way, for example, concerning how to think about—what intellectual tools we need to make sense of—things that can be quantified, divided, measured, etc.
2. The aspect-revelation in the proof regarding *x*/0 happens differently than the way the duck reveals itself in ; these are different kinds of aspect-experiences. In the duck–rabbit case, a concept—a certain possible determination—is guiding our perception, yet it does not force a conceptualization on us, compel us to judge that  is a duck-picture. The mathematical proof, on the other hand, leads us to a position where we are no longer guided by any particular concept; we are there left guided by our raw, "primitive," preconceptual, pre-mathematical, sense of things. This sense, however, compels us to endorse a certain determination: to impose certain limits on what 'divide' can mean.
3. The idea that proofs reveal aspects should be separated from the idea that proofs prepare or establish practices. The ideas are related but not interchangeable. I emphasize this, because it is hard to see it in Wittgenstein, who sometimes even uses the first idea to suggest the second, calling attention to the way proofs trigger aspect-shifts to bring home the idea that proofs alter or establish practices (see *RFM*, I, appendix 2; III, §37, §46, §47).[13] The distinction I'm emphasizing is also not easy to see in Juliet Floyd's writings about aspects and mathematics, to which I'm hugely indebted (although see Floyd and

[13] The aspect-experience in mathematics lacks importance, Wittgenstein says, "so long as it does not have certain consequences" (*RFM*, III, §47). He means *practical* consequences: the aspect-shift "only becomes important when it connects with the application of [a] sign" (ibid.). For Wittgenstein, that is, *in mathematics* the aspect-experience matters only insofar as it informs practice formation and determination: mathematics is preparatory (things are different, e.g., in aesthetics). Nevertheless, there is clearly a distinction for Wittgenstein: practice-determination can happen without aspect-experience: "I want to say: 'It need never have *struck* anyone that in certain products all the factors are equal'" (ibid.). Inversely, aspect-experiences can occur without practice-determination: The very idea of negative numbers, for example, may initially sound like a joke (see discussion below). This would indicate an aspect-experience even as no practice is being determined (yet), and it should not prevent one from learning to handle negative numbers.

> Mühlhölzer 2020, 204–13). But the distinction is important. To reveal an aspect is not yet to establish a practice. True, revealing aspects can be done, and in mathematics is normally done, with a view to establishing or revising practices, as with a proof that $x/0$ is undefined, or a proof of the commutative law for multiplication, after which we conduct our calculations somewhat differently. Mathematics is *preparatory*. Furthermore, in mathematics, seeing an aspect is often *psychologically* sufficient to establish or revise a practice. The mathematical aspect-experience may be enough to entrench us in the practice, so to speak, and cement our conduct. Upon seeing a compelling proof, we normally see no alternative. And yet, even when psychologically sufficient, revealing an aspect and establishing a practice are logically distinct. Seeing the duck in , for instance, is not enough to convince us to henceforth use it as a duck-picture; seeing a weeping willow as sad does not mean that we can now unproblematically talk about the emotions of trees. This last example shows how some aspect-seeing is not at all preparatory—not anyway of any rule-governed activity; the duck–rabbit example shows that even when preparatory, the aspect-seeing may not be sufficient to fix a rule-governed practice. Both examples show that revealing an aspect and establishing a practice are distinct.

To sum up, mathematical proofs, when followed not merely algorithmically, drive us to a reflective position from which we may overview our ontological sense of, and practices regarding, quantities, magnitudes, operations, shapes, or whatever mathematics treats of. Proofs thus orient us towards things, and give us a sense of necessity about being so oriented—about having things in our world in a certain way—which may then be cemented and inscribed in mathematical laws.

Proofs and Pictures

I would like to pause, and distinguish the view I presented from a different possible view about the relevancy of aspects to mathematics. There is a debate in the philosophy of mathematics about whether pictures really can be proofs, like these pictures that prove the commutative law for multiplication, and the Pythagorean Theorem:

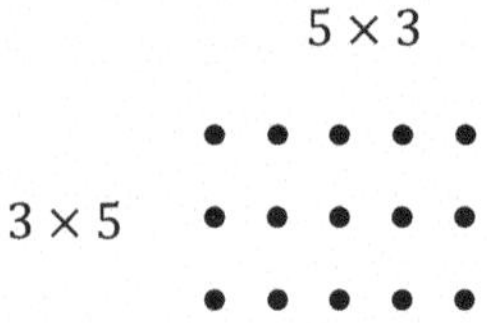

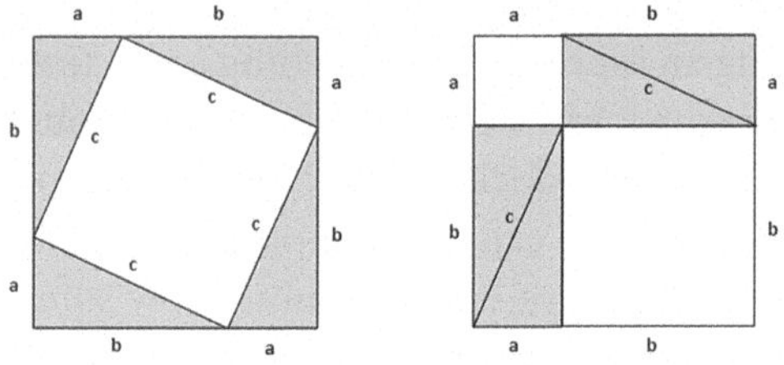

Such pictures can nonverbally alter the way we think about things in the manner proofs do. James Robert Brown furthermore thinks that "*pictures are crucial*" (1997, 165, his italics). This reacts against a prevalent attitude among mathematicians towards pictures, according to which only verbal- and formulae-proofs should strictly be considered proofs. One worry is that pictures can mislead, make us see things that are not really there: constructions that seem to trisect an arbitrary angle, for example, square a circle, or double a cube, can convince of falsehoods. We know that those things are impossible; we have proofs—algebraic proofs. (For a review of the debate see Mancosu 2005.)[14]

[14] In the background of the debate about pictures in mathematics, there is the wish to know which language is superior—the picture-language or the algebraic-language. And this wish, I suspect, is misguided; at least insofar as it harbors a fantasy about a language in which nothing misleading can be formulated. True, pictures can mislead, but so can propositions and formulae. And this is also true: we have more tools like Cartesian analytic geometry for translating pictures into formulae than for translating in the other direction; but this can hardly be a criterion for absolute linguistic superiority. A more important question, I think, is: which language gives us the kind of understanding we need? And plausibly, in different cases the answers will be different.

Now, however this debate is resolved, I believe it should not affect the discussion about aspects in mathematics. True, pictures like the ones above adduce the kind of aspect-shift proofs do. But aspect-shifting does not happen only with pictures. Aspect-shifting can happen with words, objects, events, relations, and whatnot. In seeing aspects, our mind's interface with the world is (re)shaped—how we think, understand, and behave; that is, not only how we see. Algebraic formulae are therefore not excluded. So not all proofs are visual, or involve pictures (latently or otherwise); if pictures are crucial, it is not in *this* way. Algebraic proofs too may reveal aspects.

Still, thinking about pictures—especially about the relations between propositions and pictures—could be useful in the present context: It reveals another important side of why aspects are relevant to mathematics, as well as the preparatory nature of mathematics more generally. Putting aside the question whether the 'picture theory' associated with Wittgenstein (*TLP*) and according to which propositions are pictures is good *as a theory*, it is nevertheless useful to look at propositions and pictures together. For one thing, it is a feature of both propositions and pictures that we can employ the same proposition or picture to represent both how things are and how they are *not; what a picture means is a function of its logical application and use*. For another, in both cases there is a similar potential for skepticism about meaning: One may look at the picture-proof of the Pythagorean Theorem above and not see how it proves anything, or at a written sentence in a language one can't read and see nothing but ink-marks. My discussion above does not offer a full philosophical account of meaning, but it can contribute to that discussion. For it seems reasonable to accept that seeing isn't merely passive receptivity—to put things in a Kantian manner—and so, for both propositions and pictures, one has to be mentally spontaneous in order to see: to see the proof or the proposition in one important, intensional, sense of 'see,' is to come to think *with* the picture or the proposition, for example, apply them. But this, I argued, is precisely what proofs may do: they show how theorems can guide thought, how to think with propositions and apply them.

Natural "Primitive" Reactions and Aspect-Seeing

If I'm right, mathematical proofs may necessitate conclusions by revealing aspects, which involves placing us in a reflective position, from which we can then examine our practices, and match them to our sense of the ontology of things. But what kind of reflection is it? Specifically, how intellectual is it? Wittgenstein says that aspect-seeing seems half experience, half thought (*PI*, 197). This, I suggest, is because aspect-seeing is a special kind of experience: a palpable experience of a particular concept, say, making a particular seeing-experience possible. The concept in aspect-seeing—more generally, the form of sensation, how the sensation is organized—doesn't remain transcendental, in the background, but is rather in the foreground. It becomes an object of direct attention, palpably experienced: "what we have here is a modified concept of *sensation*" (209). Now, I would like to argue that the reflection that mathematical proofs provoke is similarly partially experiential, that is, not purely intellectual and abstract. More specifically, to experience an aspect, and by extension, to understand a proof (more than merely algorithmically), is to experience a primitive reaction to something, a pre-normative sense of it. The idea of primitive reactions is also Wittgenstein's (*PI*, §244, 200, 218; Z, §541, §545; *CV*, 36), so let me explain that first. I'll then connect it to aspect-seeing and give a mathematical example.

Consider the fact that we naturally classify different shades of blue as of the same color (see *BB*, 130ff; also *PI,* §444; Z, §331, §380; WCL, 99ff.). We may ask: What justifies this? It is possible to imagine people who see not 7 colors in the rainbow, but 112, or 3; there are in fact people who divide the color-spectrum differently.[15] So why divide the color-spectrum as we do? To answer, we can appeal to physiological evolutionary or cultural factors. But instead, we can see something important if we turn the tables, and take what we naturally, that is, prelinguistically, do (whether by first or second nature, by nature or by nurture) as *explanans* rather than *explanandum*: Our classifying the shades together is "primitive"—not necessarily in the sense that it is not shaped by training, but in the sense that

[15] See Ratner (1989) for discussion.

it antecedes our concept 'blue' (Wittgenstein calls primitive reactions *pre-linguistic*; *RPP1*, §916). As such it also explains why our concept 'blue' is as it is: The reaction animates the conceptual practice and shapes it. Our natural reaction in dividing the color-spectrum as we do is one possible human reaction. Alternative reactions are possible, and so are other ways of classifying colors; but built and shaped as we are (whether by nature, by nurture, or both), those alternatives would be 'dead' to us, foreign, and unnatural. At least until we cotton onto the alternative color-language, if we ever do, we would be able to operate with such alternative color-concepts only 'algorithmically,' if at all. On the other hand, our linguistic practice with color words is an expression of our reaction, an acculturation of it: "Language [...] is a refinement" (Wittgenstein *CV*, 36; see also *PI*, §257; Z, §545). Generally, alongside our sense of the ontology of things, our primitive, natural, reactions shape how we organize things into our world, possibly conceptualize them, and they guide possible developments of our practice.[16] These reactions are a function of how we are built, also our upbringing and human form of life, and therefore—like the ontology of things themselves—quite contingent. Our orientation to things is not precalculated, or calculable: "It is not mathematically determined what is the natural thing to do" (Wittgenstein *LFM*, 187, see also *OC*, §475). Thus, at the bottom of our practices and conventions, lies something contingent. In Stanley Cavell's words, "[u]nderlying the tyranny of convention is the tyranny of nature" (1979, 123).[17]

[16] Alan Turing makes a similar point about logical contradictions: "[...] if one teaches people to carry out orders of the form '*p* and not-*q*' then the most natural thing to do when ordered '*p* and not-*p*' is to be dissatisfied with anything which is done" (in *LFM*, 187). Which is to say: one *would* be able to do something with the order; but nothing satisfying. What is meaningful and satisfying, and thus what is ultimately accept as our practice with those expressions, is thus grounded in what is contingent, contingently natural. Turing's remark, made during Wittgenstein's 1938 lectures on the philosophy of mathematics, suggests that Turing was at least sympathetic to Wittgenstein's view on this central point.

[17] It might be worth emphasizing that the 'nature' spoken of here is not what scientists are authoritative on, but is given prescientifically. Also, and relatedly, the discussion suggests the possibility of mathematically deviant people: people with different natural reactions—the wood-sellers in Wittgenstein's famous example (*RFM*, I, §149), or the even more famous deviant pupil (*PI*, §185). In general, it is difficult for us to describe such people to ourselves, and although I don't have the space to develop this, I suggest that the difficulty of describing them is one of imagining their alternative conduct as a *live* conduct. That is, we *can* describe it, but only algorithmically so to speak. Whether we can do more largely depends on us—on our nature, but also on our willingness and imaginativeness—and is not something we can know *a priori*. See Stroud (1965), Cerbone (2000), Miguens (2020). Also compare William James' discussion of 'live' and 'dead' hypotheses and options (1896).

How does this connect to aspect-seeing?—Our natural reactions, I said, animate our practices, but as such, they typically go unnoticed. Through them we see and make sense of things, but again, like a flashlight, they themselves are usually not objects of focal attention. Nevertheless, it is possible to bring them into view, for instance, by imagining people with different natural reactions. I did this above when I mentioned people who divide the color-spectrum differently than we do. Thinking about such people would typically involve an experience—a somewhat disorienting, reflective, experience—which is at least akin to aspect-experience. For a similar effect, we may compare the mathematical ontology of apples with that of clouds or shadows: it would often be nonsensical to ask: 'How many clouds are there in the sky right now?', and Wittgenstein asks rhetorically: "When my shadow coincides with Smythies's shadow, how many shadows are there…?" (*LFM*, 164). Such contrasts, i.e. between the ontology of books and that of shadows, palpably manifest how our minds are primitively aligned differently to different things—pre-practically orientated to them, and invite different conceptualizations.[18]

Wittgenstein uses the word "reaction" and its cognates ambiguously sometimes. He talks, for instance, of a "consensus of action: a consensus of doing the same thing, reacting in the same way" (*LFM*, 184). Reacting here may just mean (1) practicing. But it may also mean (2) having a similar pre-practical reaction, a similar primitive sense of the matter at hand, even before we act, and before a practice has been established. If the point of teaching mathematics is merely to get people to *act* in certain ways, then by and large it does not need to involve aspect-seeing. To recall, there is a distinction between revealing aspects and establishing

[18] We may try to imagine here not just alternative natural reactions or anthropological regularities, but also alternative physical regularities: "[i]f anyone believes that certain concepts are absolutely the correct ones, and that having different ones would mean not realizing something that we realize—then let him imagine certain very general facts of nature to be different from what we are used to, and the formation of concepts different from the usual ones will become intelligible to him" (Wittgenstein *PI*, §142, p. 230; also *LFM*, 149; *RFM*, VII, 1; *Z*, §351). If enough objects like potatoes and spoons regularly changed their shapes, for example, like clouds and fire-flames, or if they often multiplied or disappeared without reason, our sense of reality concerning shape and number would be different, and so the life of our mathematical practices. Imagining such things gives us something that is at least akin to an aspect-experience: a disorienting experience, which allows us to appreciate the contingent foundations of our mind's relation to the world.

practices (see above). One may learn and follow the commutative law for multiplication, for instance—that is, *not* merely algorithmically—yet without ever seeing the proof or experiencing this law as the right one. Nevertheless, our nature—here, our natural sensitivities and ontological sense of things—is present whether or not we reflectively experience it. Eliciting it, getting people to experience it, giving them an aspect-experience—for example, by leading them through a proof—also gives them a sense of the necessity of the practice: it impresses their own nature upon them, making their reactions' role in animating the practices palpable to them. It is a way of getting people to commit to a practice, own it, and see the practice as expressing *them*.

My discussion of impossibility proofs above demonstrates this last point about natural reactions. Here is another simpler example: When teaching a child arithmetic, we arrive at this sort of question: 'If you have 2 apples and I take 3, how many do you have left?' The child might ask for the answer, and then simply accept it and practice accordingly. When I introduced this question to my four-year-old daughter, she was baffled; she might have laughed too—reacted as if the question was a grammatical joke.[19] This would express an aspect-experience: a reflective experience of her inclination. "[T]o attempt to take [a number] away from a number less than itself is ridiculous" (Frend 1796, x). With training, what's natural can change.[20]

In higher mathematics, but even when it comes to the simplest transformations in practice—like the change after accepting negative numbers, the commutative law for multiplication, or the transition from counting to adding to determine a quantity of objects—accepting practical shifts is often not *logically* forced on us, but is rather a question of

---

[19] Compare: "A serious and good philosophical work could be written that would consist entirely of *jokes*" (Malcolm 2001, 27–8).

[20] Wittgenstein makes a parallel point about imaginary numbers: "[B]y surrounding $\sqrt{-1}$ by talk about vectors, it sounds quite natural to talk of a thing whose square is -1. That which at first seemed out of the question, if you surround it by the right kind of intermediate cases, becomes the most natural thing possible" (*LFM*, 226). In general, developments in mathematical practice may result from gaining clarity regarding our natural reactions, but also from them being altered shaped or calibrated with those of others. Relatedly, Juliet Floyd writes: "we are not mathematically omniscient" (Floyd 2000, 243). This, I take it, is at least partly because we are not psychologically omniscient, and don't *a priori* know how it would be most natural for us to react—for example, before a proof puts us in a position to fully appreciate the consequences.

whether the transition agrees with, or captures the shape of, our natural sensitivity. This point has been made by Wittgensteinians in the past (originally, I think, in Diamond 1991). What I'm emphasizing is that this is why teaching mathematics, like conducting mathematical research, often involves generating aspect-experiences: tapping into people's (perhaps our own) rudimentary, that is, prelinguistic, pre-mathematical, "primitive," sensitivity to things, calling reflective attention to how things make sense, and inviting intentional imaginative participation in shaping natural reactions.

Between Realism and Antirealism: The Sources of Mathematical Normativity

I pointed out the role of our natural reactions as a source of necessity: the role of facts in preparing our mathematical practices and grounding concepts. Have I thereby been advancing mathematical realism?

Roughly, the realist thinks that it is a fact that we cannot trisect angles, or that 2 + 2 = 4, etc. These are abstract facts—neither physical, nor mental, but still facts; and they are necessary, not contingent. Such, at least, is a typical use of 'realism.' The Wittgensteinian position I propose is different: "The [mathematical] proposition *is grounded* [...] in the physical and psychological facts that make the technique *possible*" (*RFM*, VII, 1). This is also the way in which, "mathematics is [...] an anthropological phenomenon" (Ibid., §33). Mathematics reflects what kind of life we may have with things, our ontological sense of them, how we may *have* them in our world (which is very different from what formalists think). This draws attention not to necessary facts, but to contingent facts—about us and the world—that make certain activities alive and others dead to us: apples and books don't merge, multiply, or change their shapes for no reason, we can cut ropes in half, we don't often confuse different numerals, we can make rulers from hard materials, we almost always agree in our calculations, and we can write numbers on pages without worrying that they might disappear without reason. It is also contingent that certain activities are alive for us and others not. Logical conclusions and theorems are necessary not because they correspond to rigid super-facts, but when the alternative would involve an activity that we could at best perform merely algorithmically, not as a living activity, like the case above of erasing the zeros on both sides of $a \times 0 = b \times 0$ that falls short of dividing, or like commanding someone to sit faster, or hear colors.

So, according to my Wittgensteinian suggestion, there are facts about us and the world, reference to which would provide a space for understanding why we determine our practices and fix our concepts as we do. Now, saying that we determine our practice sounds idealist, and the view is indeed somewhat similar to antirealism; specifically, it involves suspicion regarding the realist notion that mathematical propositions are about mathematical facts in the same way that ornithology is about birds (see Diamond 1996).[21] But my Wittgensteinian suggestion is again different, for instance, from the sort of conventionalism Dummett finds in Wittgenstein (Dummett 1959, 1978), according to which every step in a calculation or proof involves a determination of what is necessary—a step that is ultimately unguided, and could have been taken in other directions.

Specifically, 'determination' in my proposal comes to something very different from what determination means to the conventionalist: Determining our practice in my proposal does not imply that there is an alternative—that is, a *live* alternative—and it does not concern likings or preferences. It is rather like saying: 'I now see that there is no live alternative,' or 'the alternative that seemed alive I now see as dead; I truly only live like *this*, not like *that*.' The determination thus involves discovering that the alternatives are *silent*, even if strictly speaking they are not logically incoherent. Discovering this is nothing less than the discovery of a necessity—of how our mind may be positioned towards things—how it interfaces with the world: the discovery that things—quantities, magnitudes, shapes, and things that have these—make such and such ontological sense to us, and not another. And such discoveries typically generate mental recalibration, reorientation, mind-shift. The discovery that the alternatives are silent, I think, gives rise to the realist tendency. That is, the realization that we come alive with things in one way but not another

[21] Even here there is a difference: The antirealist typically takes the realist idea to be false, whereas the Wittgensteinian position I'm recommending is rather suspicious about the very meaningfulness of the realist idea.

feels somewhat like hitting upon a rigid super-fact; that is, it is naturally describable as such.[22] But the fact that some way of acting is the only (live) option for us, if we remember all those contingent facts about us and the world, can now be seen to be determined not by some super-fact, but by reference to a host of contingent facts such as I mentioned above. Making the determination is thus our way of owning our place in the contingent world. True, it could be misleading to say that we consciously take into account those facts about us and the world I've mentioned when we determine the practice; the determination is not typically a result of such 'calculation' or deliberation.[23] Rather, sensitivity to those facts, remaining in touch with them, even if unawaredly and unreflectively, is what gives the sense that what has been determined as our practice can indeed be *our* practice—be alive for *us*. The determination of what is mathematically necessary thus does not happen unguided, indiscriminately or capriciously (as in Dummett's Wittgenstein), but rather by reference to what is natural to us: by our *owning* what is natural—owning our own nature.[24]

Lastly, I would like to highlight where the discussion connects to aspect-perception. Wittgenstein makes a further point concerning such determinations when, in discussing rule-following and the question of determining the next number in an arithmetical series, he says: "It would be less confusing to call it an act of decision [than of discovery], though this too is misleading, for nothing like an act of decision must take place, but possibly an act of writing or speaking" (*BB*, 143)—Wittgenstein is here placing his view between discovery (realism) and decision (idealism), and, as he often does, he emphasizes the owning of an activity *in practice*. The determination, that is, is made not as a separate act, but is reflected in—we might even say it is a formal aspect of—the practicing itself: in routinely developing a series, calculating, drawing inferences, etc. Now, *in* the routine, no aspect-experience will occur. Routines are not the sort of space where aspect-seeing happens. Again, the determination Wittgenstein

[22] When so described, for example, in Platonist realism, the description is at best figurative.

[23] This connects to the fact that mathematical propositions are—that is, function as—rules, not experiential propositions: "mathematics as such is always measure, not thing measured" (*RFM*, III, 75). The truth of rules is not determined in the way that experiential propositions are.

[24] For more on Wittgenstein's position between realism and antirealism, see Gerrard (2018).

speaks of here will not be a separate act of decision in addition to the practicing, taken from the outside, but rather an embedded facet of the practicing itself, its 'form,' to use an Aristotelian notion. The determination here takes a practical nonreflective form; it is *shown*, not *said*.

However, and this is where the issue connects to aspect-seeing, sometimes, something like a separate act of decision may nevertheless take place. And this involves something that is at least akin to aspect-experiencing: pausing to consider our practice from a reflective position, possibly owning and reaffirming the practice, sometimes revising it. We can thus talk of determination of the practice that is not *in* the practice, but in reflection on it. To recall, aspect-seeing often involves a view of our practices from a reflective position, 'outside' our practices, possibly a reorienting experience; it typically involves a not-purely-intellectual, partly experiential, reflection about our concepts, not just with them. My claim now is that the determination of our mathematical practices can take a similar reflective form; it can be done, so to speak, from the sidelines of the language-game—from 'sideways-on' (if I may recycle John McDowell's phrase and apply it noncritically). Adopting such a reflective attitude may be preparatory: help us to go back to nonreflective acceptance—that is, to acceptance *in* practice. Recognizing the possibility of such reflective thinking—call it 'aspect-thinking'—can also make our role in determining our practices clearer, as well as our responsibility for them: It makes it more palpable how, and in what sense, it is *we* who determine our practices.[25]

Bibliography

Agam-Segal, R. 2012a. How to Investigate the Grammar of Aspect-Perception: A Question in Wittgensteinian Method. *Essays in Philosophy* 13 (1): 85–105.

———. 2012b. Reflecting on Language from 'Sideways-On': Preparatory and Non-Preparatory Aspects-Seeing. *Journal for the History of Analytical Philosophy* 1 (6): 1–17.

[25] For their comments and help, I'm grateful to Maria Balaska, Sorin Bangu, Avner Baz, Kevin Cahill, Ed Dain, Cora Diamond, Juliet Floyd, Duncan Richter, and Dafi Agam-Segal.

———. 2014. When Language Gives Out: Conceptualization, and Aspect-Seeing as a Form of Judgment. *Metaphilosophy* 45 (1): 41–68.

———. 2015. Aspect-Perception as a Philosophical Method. *Nordic Wittgenstein Review* 4 (1): 93–121.

———. 2018. Moral Thought in Wittgenstein: Clarity and Changes of Attitude. In *Wittgenstein's Moral Thought*, ed. R. Agam-Segal and E. Dain, 67–96. New York: Routledge.

———. 2023. Avner Baz on Aspects and Concepts: A Critique. *Inquiry* 66(3): 417–449.

Bangu, S. 2019. Hard and Blind: On Wittgenstein's Genealogical View of Logical Necessity. *Philosophy and Phenomenological Research* 102: 1–20.

Baz, A. 2000. What's the Point of Seeing Aspects? *Philosophical Investigations* 23 (2): 97–121.

———. 2019. Bringing the Phenomenal World into View. In *Wittgenstein on Objectivity, Intuition, and Meaning*, ed. J. Conant and S. Greve, 100–118. Cambridge: Cambridge University Press.

Brown, J.R. 1997. Proofs and Pictures. *The British Journal for the Philosophy of Science* 48: 161–180.

Cantor, G. 1996 [1874]. On a Property of the Set of Real Algebraic Numbers. In *From Kant to Hilbert: A Source Book in the Foundations of Mathematics*, Vol. 2, ed. W. Ewald, 839–843). Oxford: Clarendon.

Cavell, S. 1979. *The Claim of Reason*. Oxford: Oxford University Press.

Cerbone, D. 2000. How to do Things with Wood: Wittgenstein, Frege and the Problem of Illogical Thought. In *The New Wittgenstein*, ed. A. Crary and R. Read, 293–314. London: Routledge.

De Villiers, M. 1990. The Role and Function of Proof in Mathematics. *Pythagoras* 24: 17–24.

Diamond, C. 1984. Review of Remarks on the Philosophy of Psychology by Ludwig Wittgenstein. *Philosophical Review* 93 (3): 458–462.

———. 1991. The Face of Necessity. In *The Realistic Spirit*, 243–266. Cambridge, MA: MIT Press.

———. 1996. Wittgenstein, Mathematics, and Ethics: Resisting the Attractions of Realism. In *The Cambridge Companion to Wittgenstein*, ed. H. Sluga and D. Stern, 226–260. *Cambridge*: Cambridge University Press.

Dummett, M. 1959. Wittgenstein's Philosophy of Mathematics. *Philosophical Review* 58: 324–348.

———. 1978. Reckonings: Wittgenstein on Mathematics. *Encounter* 50 (3): 63–68.

Floyd, J. 2000. Wittgenstein, Mathematics and Philosophy. In *The New Wittgenstein*, ed. A. Crary and R. Read, 232–261. London: Routledge.

———. 2010. On Being Surprised: Wittgenstein on Aspect-Perception, Logic, and Mathematics. In *Seeing Wittgenstein Anew*, ed. W. Day and V. Krebs, 314–337. Cambridge: Cambridge University Press.

———. 2018. Aspects of Aspects. In *The Cambridge Companion to Wittgenstein*, ed. H. Sluga and D.G. Stern, 2nd ed., 361–388. Cambridge: Cambridge University Press.

Floyd, J., and F. Mühlhölzer. 2020. *Wittgenstein's Annotations to Hardy's Course of Pure Mathematics*. Cham: Springer.

Frend, W. 1796. *The Principles of Algebra*. London: G. G. and J. Robinson.

Gerrard, S. 2018. A Philosophy of Mathematics Between Two Camps. In *The Cambridge Companion to Wittgenstein*, ed. H. Sluga and D.G. Stern, 2nd ed., 152–179. Cambridge: Cambridge University Press.

James, W. 2015 [1896]. The Will to Believe. In *The Will to Believe and Other Essays in Popular Philosophy*, 1–31. New York: Dover.

Knott, H. 2017. On Reinstating "Part I" and "Part II" to Wittgenstein's Philosophical Investigations. *Philosophical Investigations* 40 (4): 329–349.

———. 2020. On Reinstating "Part I" and "Part II" to Wittgenstein's Philosophical Investigations—A Supplementary Note. *Philosophical Investigations* 43 (4): 382–390.

Malcolm, N. 2001. *Ludwig Wittgenstein: A Memoir*. Oxford: Clarendon.

Mancosu, P. 2005. Visualization in Logic and Mathematics. In *Visualization, Explanation and Reasoning Styles in Mathematics*, ed. P. Mancosu, K.F. Jørgensen, and S.A. Pedersen, 13–30. Dordrecht: Springer.

Miguens, S., ed. 2020. *The Logical Alien: Conant and His Critics*. Cambridge: Harvard University Press.

Mühlhölzer, F. 2005. 'A Mathematical Proof Must Be Surveyable' What Wittgenstein Meant by This and What It Implies. *Grazer Philosophische Studien* 71: 57–86.

Mulhall, S. 1990. *On Being in the World: Wittgenstein and Heidegger on Seeing Aspects*. London: Routledge.

Ratner, C. 1989. A Sociohistorical Critique of Naturalistic Theories of Color Perception. *The Journal of Mind and Behavior* 10 (4): 361–372.

Strawson, P. 1974. Imagination and Perception. In *Freedom and Resentment and Other Essays*, 50–72. London: Methuen.

Stroud, B. 1965. Wittgenstein and Logical Necessity. *Philosophical Review* 74 (4): 504–518.

Travis, C. 2016. A Room in a View. In *Wollheim, Wittgenstein, and Pictorial Representation*, ed. G. Kemp and G.M. Mras, 3–33. London: Routledge.

Weil, S. 1970. *First and Last Notebooks*. Trans. R. Rees. London: Oxford University Press.

Wittgenstein, L. 1922. *Tractatus Logico-Philosophicus*. Trans. C.K. Ogden. London: Routledge. (TLP)

———. 1953. *Philosophical Investigations*. Ed. G.E.M. Anscombe and R. Rhees, Trans. G.E.M. Anscombe. Oxford: Blackwell. (PI)

———. 1958. *The Blue and the Brown Books*. Oxford: Blackwell. (BB)

———. 1969. *On Certainty*. Ed. G.E.M. Anscombe and G.H. von Wright, Trans. G.E.M. Anscombe and D. Paul. Oxford: Blackwell. (OC)

———. 1976. *Wittgenstein's Lectures on the Foundations of Mathematics: Cambridge 1939*. Ed. C. Diamond. Ithaca, NY: Cornell University Press. (LFM)

———. 1978. *Remarks on the Foundations of Mathematics*. Ed. G.H. von Wright, R. Rhees, and G.E.M. Anscombe, Trans. G.E.M. Anscombe. Oxford: Blackwell. (RFM)

———. 1980a. *Remarks on the Philosophy of Psychology*, Vol. 1. Ed. G.E.M. Anscombe and G.H. von Wright, Trans. G.E.M. Anscombe. Oxford: Blackwell. (RPP1)

———. 1980b. *Remarks on the Philosophy of Psychology*, Vol. 2. Ed. G.H. von Wright and H. Nyman, Trans. C.G. Luckhardt and M.A.E. Aue. Oxford: Blackwell. (RPP2)

———. 1981. *Zettel*. Ed. G.E.M. Anscombe and G.H. von Wright, G.E.M. Anscombe. Oxford: Blackwell. (Z)

———. 1990. *Last Writings on the Philosophy of Psychology*: Vol. 1. Ed. G.H. von Wright and Heikki Nyman, Trans. C.G. Luckhardt and M.A.E. Aue. Oxford: Blackwell. (LW1)

———. 1992. *Last Writings on the Philosophy of Psychology*: Vol. 2. Ed. G.H. von Wright and Heikki Nyman, C.G. Luckhardt and Trans. M.A.E. Aue. Oxford: Blackwell. (LW2)

———. 1993. Philosophy. In *Philosophical Occasions 1912–1951*, ed. J.C. Klagge and A. Nordmann, 158–199. Indianapolis: Hackett. (PO)

———. 1998. *Culture and Value*. Ed. G.H. von Wright, H. Nyman and A. Pichler, Trans. P. Winch. Oxford: Blackwell. (CV)

———. 2009. *Philosophical Investigations*. Ed. P.M.S. Hacker and J. Schulte, Trans. G.E.M. Anscombe, P.M.S. Hacker and J. Schulte. Oxford: Blackwell. (PI)

———. 2016. *Lectures, Cambridge 1930–1933, From the Notes of G. E. Moore.* Ed. D. Stern, B. Rogers, and G. Citron. Cambridge: Cambridge University Press. (LC)

———. 2017. *Wittgenstein's Whewell's Court Lectures*. Ed. V.A. Munz and B. Ritter. Malden: Wiley Blackwell. (WCL)

Wollheim, R. 1980. *Art and Its Objects*. Cambridge: Cambridge University Press.

Zwicky, J. 2010. Mathematical Analogy and Metaphorical Insight. *Journal for the Learning of Mathematics* 30 (1): 9–14.

———. 2014. *Wisdom and Metaphor*. 2nd ed. Edmonton: Brush Education.

Entering a Practice of Mathematical Proof: On the Difficulty of Teaching Proofs

Kim-Erik Berts

Introduction

Mathematical proof is a concept that attracts much attention in the philosophy of mathematics as well as in the didactics of mathematics. This shared interest is a natural consequence of the great significance of proof within mathematics. This significance is reflected in a widespread consensus concerning the importance of teaching proofs and including proof-like activities in mathematics education, already at an early stage of schooling.

However, it is widely recognized that proofs are difficult for students to grasp. The challenges for students and the, sometimes, meagre results of teaching proofs provide a strong incentive for teachers to omit proofs and instead focus on other activities that—presumably—benefit their students more. During the last decades, an extensive discussion has emerged about the roles of proofs in teaching. Many of these studies offer

K.-E. Berts (✉)
Åbo Akademi University, Vasa, Finland
e-mail: kim.berts@abo.fi

K. M. Cahill (ed.), *Wittgenstein on Practice*,
https://doi.org/10.1007/978-3-031-68656-6_10

explanations as to why students experience difficulties with proofs and suggestions as to how these difficulties can be alleviated. While there are valuable insights to be gathered from that discussion, I shall argue that there are further insights to be gained by incorporating a perspective on mathematical practice found in Wittgenstein's philosophy.

To be sure, much of the discussion on teaching proofs is attentive to mathematical practice and bases its results on empirical studies on teaching. However, Wittgenstein's emphasis on seeing mathematics as a practice and his ideas on the context-sensitivity of meaning provide a perspective on the difficulties of teaching proofs and on the choice of proofs that has not yet been fully appreciated.

This chapter will first review some of the arguments—from an educational as well as from a philosophical point of view—for including proofs and proof-like activities in teaching and the difficulties of teaching proofs that has been documented in research. It will then review some of the central points from the discussion on handling this difficulty and proceed to argue that a greater context-sensitivity can further the understanding of the difficulties. This chapter stands (as does its author) with one leg in the philosophy of mathematics and one leg in mathematics education; it builds on the ideas of both fields, and the ideas presented can hopefully be of use in both.

Arguments for Including Proofs in Teaching

Proof is one of the central concepts in mathematics and one of the most important activities undertaken when doing mathematics. Today, the inclusion of proofs in teaching is generally recommended in research on mathematics education following a renewed interest in the topic in the last decades (Mariotti 2006). The interest in proof in mathematics education has varied considerably over the years. Traditionally, proofs have mainly been restricted to the teaching of Euclidean geometry. The new math movement in curriculum reform in the 1960s led to an increased emphasis on proof and deduction. The 1970s and 1980s witnessed a backlash against the new math and a turn back to the basics: to routine calculations and a greater reliance on intuitive arguments (Hanna and

Jahnke 1993; Hanna and Knipping 2020). The current renewed interest emphasizes that educators would do well to include proofs or proof-like activities not only in geometry but in the teaching of other branches of mathematics as well.

In contemporary education research, the reasons for including proofs often hinge on a set of mathematical competencies or skills that studies in mathematics can foster. These are considered valuable within mathematics but possibly transferable into other contexts. With regard to developing skills such as creativity and logical reasoning, activities such as producing and trying to understand proofs are particularly well suited. The following quote from Heinze and Reiss (2009, 192) may serve as an example of this line of reasoning:

> Developing reasoning and proof competency is a significant objective of mathematics instruction at school. This competency includes the ability to reason coherently and systematically, to identify valid mathematical arguments, and to establish a correct mathematical proof.

Another reason to include proofs departs from the observation that it would be valuable to give pupils and students an insight into the practice of the science of mathematics, in so far as this is possible. Since proof is the central method for providing justification for results in mathematics, teaching proofs is necessary.

The answers that a teacher gives to questions such as "Which proofs?", "When?", and "How?" will to a great extent determine whether the learning process turns out to be fruitful or not. These questions have been discussed extensively during the last decades and I will not dwell on them here but return below to some recommendations given, in so far as they have a bearing on the theme of this chapter. I will instead review two philosophical points concerning proofs that have direct implications for the teaching of proofs.

In the philosophy of mathematics, the interest in mathematical proofs is tied to the desire for a greater understanding of the concept itself and forms a part of larger inquiries into the nature of mathematical knowledge or the foundations of mathematics. In philosophy, there is a tension between, on the one hand, the perspective of mathematical logic where a proof is defined as a formal derivation (see, e.g. Shoenfield 1967, 5), and, on the other hand, the perspective of mathematical practice where the activity with proofs is taken as the point of departure for the investigations. A prominent feature of the tradition that stresses mathematical practice is that proofs are viewed as having a greater importance than merely providing justification for theorems. Rav (1999, 20) expresses this:

> *[P]roofs rather than the statement-form of theorems are the bearers of mathematical knowledge.* Theorems are in a sense just tags, labels for proofs, summaries of information, headlines of news, editorial devices. The whole arsenal of mathematical methodologies, concepts, strategies and techniques for solving problems, the establishment of interconnections between theories, the systematisation of results – the entire mathematical know-how is embedded in proofs.

If proofs, by contrast, are reduced to formal derivations (which in practice will be too long for any human being to overview), the crucial aspects of proofs that Rav stresses will be excluded. Thus, from Rav's view we can distil a strong argument for including proofs in teaching. To put it bluntly, if proofs are excluded "the entire mathematical know-how" would be bypassed. At this point, there seems to be an evident counterargument to Rav's view. The fact that proofs figure rather sparsely in much of teaching up to secondary level entails that either Rav is wrong, or we would have to conclude that these students have, in fact, not learnt much mathematics. Since the latter conclusion seems somewhat drastic, it remains to reconsider Rav's remark. So much can be said in defence of Rav that there could probably not be a university level programme in mathematics that bypassed proof. At the elementary and secondary levels, however, there may be other ways of conveying the necessary know-how even if it does not provide the level of depth that proofs do. I shall leave that discussion, however, and be content to note that Rav indeed provides a strong argument for including proofs in teaching.

The second argument for including proofs that I will review here is a consequence of Wittgenstein's writings. This is the idea that the proof is constitutive for the meaning of the theorem (of which it is a proof). Wittgenstein's idea is related to Rav's but can be described as more radical since not only the know-how involved, but also the meaning of the theorem, is shaped by the proof. In an oft-quoted remark he states: "If you want to know what is proved, look at the proof" (PG, p. 369). In his later writings, this line of thought was tied to the idea that understanding a proof involves learning a new concept or a new criterion for something (RFM, III § 45, VII § 72; cf. Berts 2016 for a discussion of this idea).[1] If we take this idea seriously, omitting proofs from teaching robs the students of the possibility to understand mathematical results. As was the case with the previous argument for teaching proofs, Wittgenstein's remarks seemingly imply that students do not understand the theorems that they are taught unless they take part of and understand proofs of them. And, again, the consequences for the learning of students who are not taught proofs seem somewhat drastic. However, at least to some degree we do say that the student who has understood a proof of a result has a deeper understanding than the student who has not. Whatever the case, I do not think it follows from Wittgenstein's remarks that understanding of mathematical results come *only* through proofs, nor that one can achieve a complete or final understanding as soon as one has understood a proof. As Dawson (2006) points out, finding a new proof of an already established theorem is valuable—not because it increases the level of verification of the theorem, but because the new proof adds to our understanding of the theorem.[2]

That similar ideas about proofs are found in mathematics education research is evident in the following quote from Schoenfeld (2009, xiv):

> [P]roof is a way to figure out how things work. If you think something might work, you try to prove it – you try to work out *why* it should be the way you think it is. Sometimes you succeed, sometimes you don't – but

[1] This view runs counter to a common idea which, as Berts (2016) argues, is tacitly presupposed in a wide array of writings on the philosophy of mathematics. This is the idea that a theorem is a proposition that is possible to understand as it stands and that the role of the proof is to show that what the theorem states is true (and ideally also *why* it is true).

[2] There is an interesting discussion about the relation between two different proofs that prove one and the same theorem. If a proof involves giving me new concepts or criteria—it seems that a proof proves a unique theorem and that a different proof of what appears to be the same theorem, in fact proves a different theorem. I will, however, bypass this discussion here, but see, for example, Shanker (1987).

> even in failure you may come to a better understanding of the phenomenon you're trying to make sense of. That is, *proving* is a form of mathematical sense making.

In addition to the arguments found in the research on mathematics education and in the philosophy of mathematics, there is an explicit requirement to include proofs or teach proving in the curricular documents of many countries. The *Principles and Standards for School Mathematics* published by the National Council of Teachers of Mathematics in the United States (NCTM 2000, 342) provides one of the best examples:

> Reasoning and proof are not special activities reserved for special times or special topics in the curriculum but should be a natural, ongoing part of classroom discussions, no matter what topic is being studied.

The National Core Curriculum of Finland does not place an equally strong emphasis on proof, but it still has its place as can be seen in the following description from the key content areas of the grades 7–9: "[The pupils] familiarise themselves with the basics of providing proof. They practise determining the truth value of propositions" (Finnish National Board of Education 2016).

The Difficulty of Proof

Although the arguments in favour of covering proofs in teaching are strong and multifaceted, it is a common experience among educators that proofs are difficult for students to grasp. This is well documented in many studies over the last decades (see, e.g. Harel and Sowder 2007 for an overview). The nature of the difficulties that students experience with proofs vary substantially. In the oft-quoted report by Fischbein and Kedem (1982), 40% of the Israeli students (aged 15–17, $n = 397$) who took part in their study were found to correctly identify a proof as being a proof. Still, only 10–15% refuted the need for further checks. Senk (1985) studied upper secondary students in the United States (n = 2699) taking part in a one-year course on geometry that included proof writing. This study asked students to write parts of or whole proofs. Results showed that about 30% of the students managed to complete three out of four proofs. Roughly as large a share did not complete a single proof.

In a more recent study, Stylianou et al. (2015) found that American undergraduate students (n = 535) experienced difficulties with proofs, particularly in the production of proofs. Thus, the difficulties are not isolated to any specific time or geographic location nor to any specific age group. Neither are the difficulties isolated to a specific sub-discipline of mathematics as Fishbein and Kedem (1982) as well as Stylianou et al. (2015) used proofs both from algebra and geometry, while Senk (1985) focused exclusively on geometry proofs.

At this point, I shall mention two intuitively plausible explanations to these difficulties, only to show that a more nuanced account is called for. We may, first, think that the students who experience difficulties with producing proofs are simply not mature enough for such an abstract task. This explanation may find support in Piaget's theory about the stages of development of the child. Mathematical-logical thinking belongs to the formal-operational stage, which is only entered sometime between ages 11 and 15 (Inhelder and Piaget 1958). However, the difficulties with proofs have been documented also among beginning university students (Stylianou et al. 2015; Recio and Godino 2001).[3]

Another explanation could be that learning proofs is difficult due to a lack of interest or motivation. However, in several studies that report on problems with the understanding or production of proof, the same students also show an awareness of the importance of proof and are interested in learning proofs. As is documented in Hemmi (2008a), 90% of the students in her study disagreed totally or partially to the statement "I see no meaning with proof; Famous mathematicians have already proved all the results". Roughly an equally large share agreed totally or partially

[3] Moreover, Piaget's theory of stages has been challenged by later empirical research. For example, Maher (2009) has documented deductive reasoning in relation to problem-solving tasks among 8-year-old pupils.

to the statement "Proof is essential to mathematics". However, this high degree of motivation did not prevent the students from experiencing problems with learning proofs. In Stylianou et al. (2015), students saw proof as an important part of mathematics teaching. Students were divided into two categories (high-performing and low-performing) and both categories agreed to statements as the following "I think constructing proofs is an important part of doing mathematics" (agree: 80% and 48%, respectively; disagree: 5% and 17%) and "It is important to me to participate in constructing mathematical arguments or proofs during instruction" (agree: 67% and 45% respectively; disagree: 11% and 22%) (Stylianou et al. 2015, 110). Although the low-performing group of students can be interpreted to show a lower degree of motivation than the high-performing group, a considerable share still agreed to the statements. In conclusion, the reason for the widespread difficulties with learning proofs must at least partially be sought elsewhere than in a lack of motivation.

The Different Roles of Proofs

I will now review a very influential discussion from mathematics education research that has emerged as an answer to the difficulties of teaching proofs. Whereas the new math stressed the formal aspects of proofs, the renewed interest in proof that emerged in the 1980s and 1990s took its lead from the role of proof in mathematical practice; Hanna and Jahnke (1993, 422) speak of: "a shift to a pragmatic view of proof". A common objection to formal proofs is that although they are rigorous and prove the theorem, they do not necessarily carry explanatory power (Hanna 1990). They may not even be convincing to the reader. This led researchers in mathematics education to, on the one hand, emphasize proofs that are accessible to students, and, on the other hand, study different roles that proofs actually have in mathematical practice.

In one of the articles that started this discussion, Bell (1976) distinguishes three different roles that a proof can play:

1. Verification or justification
2. Illumination (or explanation)
3. Systematization

de Villiers (1990) adds two:

4. Facilitating new discoveries
5. Communication

Other additions have been made and in Hanna (2000) we find the following:

6. Construction of an empirical theory
7. Exploration of the meaning of a definition or the consequences of an assumption
8. Incorporation of a well-known fact into a new framework

Not every addition will be mentioned here but Hemmi (2008a) adds the following that, although probably not unique to proofs, echoes one of the arguments for teaching proofs:

9. Transfer (of techniques and skills from work on a proof to other situations or proofs)

These distinctions are the results of a philosophical or conceptual effort, and several of them are reflected in Rav's characterization of proofs. They allow for a continued emphasis on including proofs in teaching even if the exaggerated use of formal proofs is criticized. A formal proof is associated mainly with *verification* whereas other kinds of proof are viewed as valuable due to the other functions. According to Hanna (1990, 2000), understanding is prior to rigour from the point of view of actual mathematical practice and she therefore advocates the inclusion of proofs that have an explanatory role, since such proofs are more beneficial to the understanding for the students than proofs that merely verify the truth of a theorem.

> [I]n the classroom, the fundamental question that proof must address is surely "why?". In the educational domain, then, it is only natural to view proof first and foremost as explanation, and in consequence to value most highly those proofs which best help to explain. (Hanna 2000, 8)

Hanna's recommendation is shared by Hersh (1993) and, more recently, by Rocha (2019).

The above discussion of the different roles of proof suggests a third putative explanation to the difficulties that students experience while learning proofs, namely that they arise as a result of a poor selection of proofs. It is conceivable, for example, that a bias towards teaching proofs that emphasize verification or justification will lead to a one-sided comprehension among the students; at least this seems to have been the case with the new math. A teacher that heeds the recommendation of Hanna could then, presumably, expect better results. Such a conclusion appears too hasty, however, since the proofs in Senk's (1985) study are not of the formal, justificatory kind, but rather traditional geometry proofs that can be described as having a strong explanatory character. Nevertheless, the students in Senk's study did not fare particularly well in producing proofs, as was seen above.

At any rate, distinguishing the different roles that a proof can play in mathematical practice has provided educators with a valuable theoretical framework for understanding what happens in a classroom and for selecting proofs for instruction. Its value lies in that it does justice to the practice of mathematics and puts the focus on aspects that are important when it comes to understanding proofs.[4]

However, approaching mathematics as a practice, and more specifically, approaching *proving as a practice* can have even more far-reaching consequences for our understanding of the concept of proof and the learning of proof. I shall now continue by expanding on this idea and show that it can give a new illumination to the issues discussed above.

[4] If identifying the different roles of proof stem from a conceptual investigation of proof, Harel and Sowder (1998) approach the issue empirically. In student's work with proofs they distinguished three stages of development. These stages, or proof schemes, concern what the students found to be convincing and are: (1) external conviction proof schemes, (2) empirical proof schemes, and (3) analytical proof schemes. In contrast to Piaget, they do not associate these stages with any particular age group.

Proof as a Practice

In her work on proof, Hemmi (2008a) has studied how students enter into the mathematical practice of a mathematics department at a university in Sweden. The theoretical frame of her work was the socio-cultural theory of Lave and Wenger (1991), where learning is seen a process of entering a community of practice, through, what they call, *legitimate peripheral participation*. To become a participant in such a practice is mediated by *artefacts*, and in the context of becoming a part of a mathematical practice, Hemmi interprets proofs as artefacts. This perspective highlights the social relations in the practice and thus widens the perspective from a narrow focus on cognitive aspects of learning.

As was mentioned above, Wittgenstein remarks that a proof is constitutive of the meaning of the theorem, and he also stresses that when we understand a proof we learn a new concept or change our views on a concept. In part VII of *Remarks on the Foundations of Mathematics*, we find an elaboration of this idea when he comments: "'To give a new concept' can only mean to introduce a new employment of a concept, a new practice" (RFM, § 70, cf. also III § 31). Hertzberg (2022) points out that, in Wittgenstein's later thought, the notion of practice is related to rule-following and the kind of normativity associated with following a rule or failing to do so.

If we view the activity of proving in mathematics as a means for determining the use of concepts and shaping our practices, this entails that proving is a rather peculiar activity. While Hemmi takes the mathematics department she studied as the community of practice and treats proof as an artefact in this practice, I will instead focus on proving itself as a practice, and practice will here be understood more broadly along the lines of Wittgenstein.

Thus, I suggest that Wittgenstein's idea could be extended in the following way: (1) Not only is mathematics a practice that we enter as we learn, proving too should be regarded as a practice (or sub-practice) since it is a specific activity with its own goals, accepted moves and designated place within mathematics. (2) Learning to prove and to understand proofs is to enter into this practice. And, (3) becoming part of this practice *consists in* learning, acquiring, a new mode of verification, which is unique in so far as the practice of proving in mathematics is unique. Viewing proof in this manner has several consequences for the learning of proofs and for how we understand the difficulties that students experience when learning proofs and to prove.

I shall begin the exploration of this perspective by reviewing some comments made in connection to empirical research on the problems that students face when learning proofs, because I think this perspective allows us to see these problems as something to be expected from learners struggling to become participants in the practice of proving—not as an odd weakness in the students' understanding. In their hypothesis, Fischbein and Kedem (1982, 128), assume that the students "do not have a clear idea of what a formal mathematical proof means" and that they "will encounter great difficulty in understanding *and effectively using* the concept of formal, universal validity". Their conclusion is that students incline toward an empirical view of the validity of the proofs.[5] However, this should perhaps not be seen as a shortcoming on their part, but rather as an indication that the practice of proving in mathematics is yet foreign to them. They, therefore, turn to a mode of verification with which they are presumably more familiar, that is, the empirical-inductive approach where a greater number of observations increases confidence in what is observed. Similarly, Harel and Sowder (2007, 824) state that "[f]or some university students, even recognizing whether a given argument constitutes a proof cannot be taken as a given". Viewed from the perspective advocated here, these findings appear as something to be expected.

The perspective advocated here also sheds light on a common predicament in the teaching of proofs: the proofs that are accessible for younger students often concern theorems, which they do not feel a need to prove. The theorems often concern properties that are of such an elementary nature that the students find proving them pointless because they seem self-evident. Schoenfeld (2009) exemplifies this with the proof that the base angles of an isosceles triangle are equal. When introducing the

[5] The students in the study of Fishbein and Kedem are thus using an empirical proof scheme, if we consider the result in light of Harel's and Sowder's (1998) taxonomy.

above-mentioned distinction between the three different roles of proof, Bell (1976, 24) writes:

> Some teachers have said that proof, for a pupil, is what brings him conviction. Although this is a valuable remark, in that it directs attention to the need for classroom explanations to have meaning for the pupil rather than be formal rituals, it is perhaps dangerous in that it avoids consideration of the real nature of proof. Conviction is normally reached by quite other means than that of following a logical proof.

In the case of students, the authority of the teacher or an argument via examples usually suffice as verification.[6] Indeed, as Schoenfeld (2009) mentions, it can be a source of frustration for students if they have to prove things that appear as self-evident to them (particularly if the proof is difficult to grasp). One of the mathematicians in Hemmi's (2008a), 205) study even commented that students had more problems with proving something that they experienced as self-evident. Thus, appealing to a need for a greater rigour or appealing to their critical examination—will not necessarily awaken a need for proof among students. Not seeing a point in proving reflects a lack of familiarity with the practice of proving, where the need for further reasons deviates from most other contexts. What is required of the students is a re-evaluation of what settles a result in mathematics; one could say that learning proofs involves learning a new way of becoming convinced. I would say that when learners take part of a proof, they are often studying it with a view to learning which inferences are possible—they are *not checking* if the inferences are correct, as one would do when reading with a critical eye. They are figuring out the characteristics of the practice of proving, and an important aspect of the attitude that is required for a successful learning is that one learns to bracket the things that appear as evident at first glance. A successful entering the practice of proving is not distinguishable from learning a new conception of evidence—one where, for example, inductively checking further examples does not suffice.

[6] That is, they hold external conviction or empirical proof schemes to speak with Harel and Sowder (1998).

Hertzberg (2022, 25) makes a distinction between practices that we can choose to participate in—like chess or football—and practices that we have no choice but to participate in—like politeness. To refrain from behaving politely is not to opt out of the practice, it is in most cases simply to be impolite. I would argue that proving is a practice that one, in an obvious sense, can choose whether one participates in or not. In another sense, however, proving shares many features with reasoning rationally and coherently in a manner that lets other people understand one's thoughts. This is a practice that does not admit of choice. This double nature of proving may shed light on the reactions to students' shortcomings as they try to understand the working of a proof or to produce one themselves. In several studies that document such difficulties, there is no mention of the teaching that preceded the testing of the students' ability to recognize or produce proofs (see, e.g. Fischbein and Kedem 1982). In others, it is explicitly stated that the students have not yet taken university courses that cover proofs (Recio and Godino 2001; Stylianou et al. 2015). The proofs used to test the students in these studies are, however, such that the students can be expected to know the mathematics required to understand them. These observations suggest a tacit assumption on part of the researchers that it is meaningful to test the students' ability to understand proofs without much explicit instruction on proving. The students are tacitly taken to have acquired the reasoning or proving skills without explicit instruction. However, even if the students may be part of the more or less ordinary practice of giving reasons and arguing coherently, it may be precipitous to assume that they can transfer these skills to the context of mathematical proof. With regard to the possibilities that a shared practice opens to us, Hertzberg (2022, 27) remarks: "Being taught a practice makes me capable of carrying out certain activities – and in another sense, the existence of the practice makes doing those things possible for me." If the students are not (yet) part of the practice of proving, it is expectable that they will fail, or revert to another kind of reasoning with which they are supposedly more familiar, for example, arguing from individual instances (using an inductive-empirical proof scheme). Thus, the problem that teachers of proof regard as a lack of understanding in their students is perhaps not a failure to comprehend something that they should be able to comprehend, but rather an absence of criteria against which the validity of the proof is to be judged. What is lacking is an understanding of what is to be counted as proof. This is reflected in Hemmi's (2008a, 204) study where one of the mathematicians at the department made a similar remark:

> I very often get the question from the students: "When is a proof finished?" or "How does one start a proof?" These two issues, I think characterise students' difficulties. Either they do not understand how to start or, after making a lot of calculations, do not know if they have proved the statement.

I will now turn to the issue of the different roles that proofs were found to play by mathematics education researchers (see above) and view them through the lens of proof as a practice. As was mentioned above, the recommendation that teachers focus on proofs that have an explanatory character can be valuable in that it reminds teachers that conviction and verification have a limited and complicated status for the students. However, for the students to be able to enter a practice of proving—a transition that includes learning a new mode of verification—to focus exclusively on proofs that explain seems too limited. It will be necessary to also include such proofs that deal with properties that seem superfluous to prove, since these may be better suited to make the kind of rigour that we find in mathematics visible. Moreover, the recommendation to focus only on explanatory proofs can also reflect a problematic assumption. Do explanatory proofs have an explanatory role in themselves—or any other of the above-mentioned roles for that matter? Does a proof play a role simply because of its formulation? If we take seriously the idea that proving is a practice we should also consider that what role a proof has depends on the context in which it is presented. With a slight alteration of the oft-quoted § 43 of *Philosophical Investigations* (PI) we might say: "the role of the proof is its use." For example, when a proof is studied by a student, who is new to proofs, and who studies the proof with the intent of learning to prove, the proof will not be explanatory in the same manner as it is to someone who studies the proof to learn more about the concepts involved, that is, to someone who is familiar with the practice of proving. It would thus be pointless to choose a certain proof (or kind of proof) and count on it being explanatory in itself, as for example,

Hanna (2000, 9) does when she mentions that most proofs in geometry are explanatory.[7] Consequently, it is up to teachers to choose, not only which proofs to include in teaching but, in particular, which aspects of the practice of proving that they want to make visible for the students.

Finally, in light of the above remark, we should distinguish between teaching proof for the sake of increasing the understanding of a theorem (which was mentioned above as one of the reasons to include proofs in teaching) and teaching proof, or proving, for the sake of learning to prove. Both of these approaches to proof have their place in mathematics teaching if we want to give students valuable learning opportunities. Still, if we want students to become participants in a practice of proving, learning to prove cannot be bypassed.

Conclusions for Teaching

If we want to facilitate the process of students entering a practice of proof, what are the implications of the above discussion? I will conclude by discussing some suggestions that can be found in the literature on teaching proofs that, in light of the perspective advanced here, seem to be worthwhile.

The first and most obvious conclusion will simply be restating what was said at the outset of this chapter, that is, that teachers need to teach proofs and proving. Although proof has been a strongly emphasized theme in mathematics teaching during the last two or three decades, this has not always had an impact on what takes place in the classroom. The university students who took part in the study by Stylianou et al. (2015) did not have much experience of proofs prior to entering tertiary

[7] Hanna (2000, 8) also states that: "Some proofs are by their nature more explanatory than others." The idea of viewing proofs as explanatory due to the characteristics of the proof itself may be a consequence of its heritage. Hanna (1990) quotes Steiner (1978, 143) who maintains that "an explanatory proof makes reference to a characterizing property of an identity or structure mentioned in the theorem, such that from the proof it is evident that the result depends on the property". As it stands, this property is seemingly something that a proof can exhibit in itself. Steiner, however, remarks that this is an attempted explication of mathematical explanation, not of "relative explanatory value". And, in teaching, it would seem that relative explanatory value is the principle guiding the choice of proof.

education. Only 49% reported that their teachers had presented proofs of results to the class. Furthermore, "only 25 % of students reported that previous classroom experiences emphasized the importance of developing proofs, and only 19 % reported that constructing proofs was a regular part of classroom activity" (p. 112).

The second point is that we need to create a learning situation that communicates that it is a valuable and expected outcome that the students take steps towards becoming a part of the practice of proving. Stylianou et al. (2015, 112) reported that among the students that had experiences of proof, "only 49 % of students agreed that their instructors expected students to understand the proofs presented to them". We could interpret this in the following way: 51% of the students got the impression that they were not expected to enter this practice. This will most likely have a strong negative impact on learning, if we view learning as becoming a participant in a practice. Stylianou et al. (2015, 118) point to strong traditions in teaching that favour individual learning and even hinder the participation in constructing proofs.

A related point concerns the way teachers and mathematicians talk and write about proofs and proving. Hemmi (2008a, 186) noted that mathematics textbooks can "cause a feeling of inadequacy among students and, thus, can lead to an experience of non-participation". Phrases like "It can easily be seen that ...", "The proof is left to the reader as an exercise", etc. are abundant in mathematics textbooks and can give an unintended message of exclusion for someone who does not easily realize or does not manage to prove what is left to the reader. Also, the fact that mathematical texts often display only the results of the mathematical work and tend to hide the thought process that led to them can have a similar effect on the newcomers. The stark contrast between proofs found in textbooks and the often frustrating process of trying to formulate proofs (a form of legitimate peripheral participation) needs to be mediated by a teacher, who can reassure the student that mistakes, guesses, tests, even feeling lost, etc. are a natural part of the thought process. In this respect, Breger (2000, 225) sees a difference between older and newer textbooks: "The know-how necessary for the understanding of the modern textbook seems to be omitted on purpose, whereas the older books try to draw one's attention to the fact that there is a know-how." It therefore becomes an important task for the teacher to show these aspects of proving.

This brings me to the third conclusion, that it is important to incorporate the reasoning competencies that students already have into mathematics instruction in order to facilitate their becoming part of the practice of proving. They need opportunities to give reasons, question, refute, etc. in mathematical contexts if they are to understand what suffices as a reason or what counts as a refutation *in mathematics*. We find several researchers calling for this kind of activity; for example, Heinze and Reiss (2009, 193) emphasize this and connect it to a favourable social context:

> The underlying idea is mainly to support students' understanding of proofs and to establish mathematical reasoning as a habit of mind. Developing this habit presupposes a classroom atmosphere that consistently supports reasoning processes in various contexts and at all grades.

The emphasis on establishing reasoning as a habit of mind can be interpreted in different ways. One way is to focus explicitly on the concept of proof, logic, and proof techniques. Cabassut et al. (2012, 182) argues for this approach and claim that the students' problems with proofs can be "overcome only when metaknowledge about proof is made a theme of mathematics teaching". Although explicit focus on metaknowledge about proof will likely be beneficial, there is a risk that the students are not included in the practice of using the techniques that are taught. From the perspective advanced here it is clear that a rule or a formulated proof technique will have no meaning outside of the context of the practice where they are used. And, although there is an established practice of using these techniques, the problem with teaching is that the students are not yet part of this practice.[8] In a discussion of rules and games, Hertzberg (2022, 23) comments:

---

[8] In the Advanced syllabus of mathematics in the Finnish General Upper Secondary Education, there used to be a course which dealt with logic and proof techniques. My experience from teaching this course multiple times is that the students experienced the same kind of problems when trying to produce proofs as is reported in the research literature on the subject, although the lessons explicitly covered proof techniques. Interestingly, the parts of the course that treated proofs have now been replaced by algorithms and programming.

> Even with formulated rules, it is important to note that the actual playing of the game is, in an important sense, logically prior to the rule formulations: the practice shows what it is to follow the rule. It would be an error to think of the rule formulation as a kind of premise, a foundation, from which the practice is derived; rather, where rules are formulated, the formulated rules are part of the practice.

Accordingly, it would be a mistake to think that the explicit focus on metaknowledge about proofs would, in itself, be enough to introduce students to proving.

In her studies, Hemmi (2008a, 2008b) makes use of the notion of *transparency*, which she takes from Lave and Wenger (1991), and which is a property of artefacts. This describes the delicate balance between the explicit focus on details in proving (where the act of proving becomes *visible*) and the use of proof to further other mathematical aims (where the proof becomes *invisible* if we use the terminology of Lave and Wenger). In the case of proofs, transparency refers to the combination of these two foci, and it can provide an insight into why a one-sided focus on explicit proof techniques has its limitations.

As is seen in the above quote from Heinze and Reiss, they see "establishing reasoning as a habit of mind" as a suitable task for all grades. Harel and Sowder (1998, 280) also stress that the development of these competencies should begin at an early age. This brings me to another interpretation of the third conclusion. Instead of focusing explicitly on logic and proof techniques, the teacher can create learning situations where there arises a need to investigate, discuss, and argue for solutions.[9] A striking example of this is found in Maher's (2009) study of children in grade 3. The children worked with a problem-solving exercise, and as is typical in this kind of set-up, the pupils worked in pairs and their attempted solutions were subsequently followed up by the teacher and then presented to the class. In relation to the investigation, the pupils developed arguments utilizing recursion and contradiction in favour of their solutions. When further prompted by the researcher, the pupils were able refine their

[9] This kind of learning situation is associated with the vast literature on problem solving and heuristics in mathematics education, and there are many suggestions as to how a teacher can arrange this kind of learning situation. It is beyond the scope of this chapter to discuss that.

arguments. As Maher (2009, 131) notes, the pupils' arguments arose out of a need to convince themselves and others that their solution was correct. Without facing explicit requirements on rigour or formality, the pupils formulated their arguments according to the needs of the situation. Their work on the problem could be regarded as a (successful) form of legitimate peripheral participation, to use term of Lave and Wenger (1991).

Creating this kind of situation mimics the situation of a research mathematician in that it provides a strong motivation for formulating convincing arguments and for making sense. What is required in terms of rigour and generality of their proof-like arguments will increase as the students mature. If they have experienced this kind activity already from early on in their schooling, however, they will have taken small steps towards entering the practice of proving.

In conclusion, if we regard proof and proving in mathematics as a practice in the way mathematics is described as a practice in Wittgenstein's writings, the difficulties that students face when learning proofs are illuminated. Much of what is described in the literature on the difficulties of learning proofs turn out to be consequences of the struggle to enter this practice, but this is easily overlooked if we view them merely as cognitive shortcomings.

References

Bell, A.W. 1976. A Study of Pupils' Proof-explanations in Mathematical Situations. *Educational Studies in Mathematics* 7: 23–40. https://doi.org/10.1007/BF00144356.

Berts, Kim-Erik. 2016. *The Certainty of Mathematics. A Philosophical Investigation.* PhD Diss. Åbo: Åbo Akademi University Press.

Breger, Herbert. 2000. Tacit Knowledge and Mathematical Progress. In *The Growth of Mathematical Knowledge*, ed. Emily Grosholz and Herbert Breger, 221–230. Dordrecht: Kluwer.

Cabassut, Richard, AnnaMarie Conner, Filyet Aslı İşçimen, Fulvia Furinghetti, Hans Niels Jahnke, and Francesca Morselli. 2012. Conceptions of Proof—In Research and Teaching. In *Proof and Proving in Mathematics Education*, ed.

Gila Hanna and Michael de Villiers, 169–190. Dordrecht: Springer. https://doi.org/10.1007/978-94-007-2129-6_7.

Dawson, John W. 2006. Why Do Mathematicians Re-prove Theorems? *Philosophia Mathematica* 14: 269–286. https://doi.org/10.1093/philmat/nkl009.

Finnish National Board of Education. 2016. *National Core Curriculum for Basic Education 2014*. Helsinki: Finnish National Board of Education.

Fischbein, Efraim, and Irith Kedem. 1982. Proof and Certitude in the Development of Mathematical Thinking. In *Proceedings of the Sixth International Conference for the Psychology of Mathematical Education*, ed. Alfred Vermandel, 128–131. Antwerp: Universitaire Instelling Antwerpen.

Hanna, Gila. 1990. Some Pedagogical Aspects of Proof. *Interchange* 21: 6–13. https://doi.org/10.1007/BF01809605.

———. 2000. Proof, Explanation and Exploration: An Overview. *Educational Studies in Mathematics* 44: 5–23. https://doi.org/10.1023/A:1012737223465.

Hanna, Gila, and Christine Knipping. 2020. Proof in Mathematics Education, 1980–2020: An Overview. *Journal of Educational Research in Mathematics* 33: 1–13. https://doi.org/10.29275/jerm.2020.08.sp.1.1.

Hanna, Gila, and H. Niels Jahnke. 1993. Proof and Application. *Educational Studies in Mathematics* 24: 421–438. https://doi.org/10.1007/BF01273374.

Harel, Guershon, and Larry Sowder. 1998. Students' Proof Schemes: Results from Exploratory Studies. In *Research in Collegiate Mathematics Education*, ed. Alan H. Schoenfeld, Jim Kaput, and Ed Dubinsky, vol. 3, 234–283. Providence, RI: American Mathematical Society.

———. 2007. Toward Comprehensive Perspectives on the Learning and Teaching of Proof. In *Second Handbook of Research on Mathematics Teaching and Learning: A Project of the National Council of Teachers of Mathematics*, ed. Frank K. Lester Jr., 805–842. Charlotte, NC: Information Age.

Heinze, Aiso, and Kristina Reiss. 2009. Developing Argumentation and Proof Competencies in the Mathematics Classroom. In *Teaching and Learning Proof Across the Grades: A K–16 Perspective*, ed. Despina A. Stylianou, Maria L. Blanton, and Eric J. Knuth, 191–203. New York: Routledge.

Hemmi, Kirsti. 2008a. *Approaching Proof in a Community of Mathematical Practice*. Saarbrücken: VDM.

———. 2008b. Students' Encounter with Proof: The Condition of Transparency. *ZDM Mathematics Education* 40: 413–426. https://doi.org/10.1007/s11858-008-0089-9.

Hersh, Reuben. 1993. Proving Is Convincing and Explaining. *Educational Studies in Mathematics* 24: 389–399. https://doi.org/10.1007/BF01273372.
Hertzberg, Lars. 2022. *Wittgenstein on Criteria and Practices.* Cambridge: Cambridge University Press. https://doi.org/10.1017/9781108946537.
Inhelder, Bärbel, and Jean Piaget. 1958. *The Growth of Logical Thinking from Childhood to Adolescence: An Essay on the Construction of Formal Operational Structures.* London: Routledge & Kegan Paul.
Lave, Jean, and Etienne Wenger. 1991. *Situated Learning: Legitimate Peripheral Participation.* Cambridge: Cambridge University Press.
Maher, Carolyn A. 2009. Children's Reasoning: Discovering the Idea of Mathematical Proof. In *Teaching and Learning Proof Across the Grades: A K–16 Perspective*, ed. Despina A. Stylianou, Maria L. Blanton, and Eric J. Knuth, 120–132. New York: Routledge.
Mariotti, Maria Alessandra. 2006. Proof and Proving in Mathematics Education. In *Handbook of Research on the Psychology of Mathematics Education: Past, Present and Future*, ed. Ángel Gutiérrez and Paulo Boero, 173–204. Rotterdam: Sense.
National Council of Teachers of Mathematics, NCTM. 2000. *Principles and Standards for School Mathematics.* Reston, VA: NCTM.
Rav, Yehuda. 1999. Why Do We Prove Theorems? *Philosophia Mathematica* 7: 5–41. https://doi.org/10.1093/philmat/7.1.5.
Recio, Angel M., and Juan D. Godino. 2001. Institutional and Personal Meanings of Mathematical Proof. *Educational Studies in Mathematics* 48: 83–99. https://doi.org/10.1023/A:1015553100103.
Rocha, Helena. 2019. Mathematical Proof: From Mathematics to School Mathematics. *Philosophical Transactions of the Royal Society A* 377. https://doi.org/10.1098/rsta.2018.0045.
Schoenfeld, Alan H. 2009. The Soul of Mathematics. In *Teaching and Learning Proof Across the Grades: A K–16 Perspective*, ed. Despina A. Stylianou, Maria L. Blanton, and Eric J. Knuth, xii–xvi. New York: Routledge.
Senk, Sharon L. 1985. How Well Do Students Write Geometry Proofs? *The. Mathematics Teacher* 78 (6, Sept.): 448–456. https://doi.org/10.5951/MT.78.6.0448.
Shanker, Stuart G. 1987. *Wittgenstein and the Turning-Point in the Philosophy of Mathematics.* Albany: State University of New York Press.
Shoenfield, Joseph R. 1967. *Mathematical Logic.* Reading, MA: Addison-Wesley.
Steiner, Mark. 1978. Mathematical Explanation. *Philosophical Studies* 34: 135–151. https://doi.org/10.1007/BF00354494.

Stylianou, Despina A., Maria L. Blanton, and Ourania Rotou. 2015. Undergraduate Students' Understanding of Proof: Relationship Between Proof Conceptions, Beliefs, and Classroom Experiences with Learning Proof. *International Journal of Research in Undergraduate Mathematics Education* 1: 91–134. https://doi.org/10.1007/s40753-015-0003-0.

de Villiers, Michael. 1990. The Role and Function of Proof in Mathematics. *Pythagoras* 24: 17–24.

Wittgenstein, Ludwig. 1974. *Philosophical Grammar*. Oxford: Blackwell.

———. 1978. *Remarks on the Foundations of Mathematics*. 3rd ed. Oxford: Blackwell.

———. 2001. *Philosophical Investigations*. 3rd ed. Oxford: Blackwell.

Part IV

Agreeable Animals

Social Ontology, Evolution, and the Foundations of Practice Theory

Mark Risjord

Introduction: A Puzzle at the Foundations of Practice Theory

Because they ground representation on social practice, practice-theoretic approaches to language presuppose a capacity for joint action. This presupposition exposes practice theory to a potential circularity. All current accounts of joint action require sophisticated mental representations, such as propositional attitudes, mind-reading, mutual knowledge, and so on. Yet, practice theory takes all such mental representations to be products of social practices. From the perspective of contemporary social

This research was supported by the Czech Science Foundation (GAČR), GX20-05180X "Inferentialism Naturalized: Norms, Meanings and Reasons in the Natural World."

M. Risjord (✉)
Emory University, Atlanta, GA, USA

University of Hradec Králové, Hradec Králové, Czechia
e-mail: mrisjor@emory.edu

K. M. Cahill (ed.), *Wittgenstein on Practice*,
https://doi.org/10.1007/978-3-031-68656-6_11

ontology, there is simply no way that practices sufficiently sophisticated to support practice theoretic accounts of representation could arise without the prior existence of human-like representational capacities. Social accounts of representation in the vein of Wittgenstein, Sellars, or Brandom, one might argue, are circular. Call this the "social ontology objection" to practice-theoretic accounts of thought and language. The object of this essay is to rebut the social ontology objection by providing an alternative, minimalist account of joint action and thereby putting practice theory on a firm social-ontological foundation.

While circularity objections to practice theory are common (see, among many examples, Harman 1968; Fodor and Lepore 2007; Koons 2021; Loeffler 2009, 2021; Hutto and Satne 2015; Satne 2021b), the idea that the foundations of practice theory problematically rest on a form of joint action or we-intentionality have only recently been thematized. In Koreň et al.'s recent collection, *Groups, Norms, and Practices: Essays on Inferentialism and Collective Intentionality* (2021), three of the contributors (Satne, Loeffler, and Koons) are explicitly worried about potential circularities involving collective intentions. However, the essays disagree about the point at which joint action or collective intentions are required. While Satne (2021b) and Loeffler (2021) argue that Brandom's score-keeping account of propositional content requires collective intentions, Koons's (2021) analysis purports to show that collective intentionality is required before any rule-governed (or normative) behavior could emerge. Different consequences follow from the point at which social ontology is required. Brandom's scorekeeping is just one proposal for understanding propositional content in practice theoretic terms. If the circularity emerges at that point, then an obvious strategy for the practice theorist would be to adopt an alternative account of propositional content. On the other hand, if the circularity arises where Koons indicates, then *any* form of practice theory will be subject to the social ontology objection.

The different possible circularities of joint action emerge because there are no generally accepted criteria for identifying something as a "joint action" or as "jointly intentional." Since "joint intentionality" and its cognates are philosophical terms of art, the description of the analysandum is already heavily theory-laden. Without prior commitment to one or another theory of social ontology, there has been no clear way to determine whether an example is a joint action or whether it is simply two agents acting with individual intentions. The tight link between examples and analyses undermines the force of the social ontology objection. The objection rests on two premises:

1. Practice theory must appeal to joint actions prior to the emergence of propositional attitudes.
2. Any analysis of joint action must appeal to propositional attitudes.

If the phenomenon of joint action were defined by a particular kind of propositional attitude, then the practice theorist could dodge the issue by simply denying premise (1). After all, practice theorists take pains *not* to invoke the kind of mental representations invoked by analyses of joint action. However, with an analysis-neutral criterion for identifying joint actions, the social ontologist can make premise (1) stick by showing that practice theoretic accounts require joint action, even if the actions are described without representational language. The Section "A Neutral Description of Joint Action" will sharpen the social ontology objection by developing a neutral criterion for circumscribing the phenomenon of joint action.

To make matters precise, it will help to have a particular practice-theoretical account in mind. The section "Inferentialism and Joint Action" will deploy Brandom's inferentialist-expressivism (Brandom 1994, 2008, 2009). The primary advantage of Brandom's view in this context is that it is articulate about the social practices that are supposed to underpin representation. Section 3 will use the neutral criterion for joint action to argue that Brandom's account appeals to joint action at the point where normative practice emerges from the practical attitudes, a point where a variety of practice theorists are also likely to be vulnerable. The social ontology objection is thus shown to be a general concern for practice theories, not just for Brandom.

If the first premise of the social ontology objection is made to stick, then the next available exit is for the practice theorist to deny the second premise. "Minimalist" or "lite" accounts of joint action are a promising avenue. Minimalist accounts, such as those proposed by Butterfill (2011,

2017), Pacherie (2013), Tollefsen (2005, 2012), Vesper et al. (2010), or Satne (2016, 2021a, 2021b), aim to understand joint action or intentionality with resources that are less cognitively demanding than the currently dominant approaches. However, they have not been designed with the needs of practice theory in mind, and none of the analyses in the current literature are entirely satisfactory. In particular, Brandom's inferentialist-expressivism puts two constraints on any analysis of joint action.

First, on Brandom's view, full-blown practical intentionality emerges, logically speaking, only after linguistic practices are in place. "Inferentialism and Joint Action" will argue that Brandom's account therefore requires joint action at the level of *prelinguistic* practices. Some accounts, while less cognitively demanding than the standard accounts, continue to invoke intentions more sophisticated than Brandom's account could support (e.g., Pacherie 2013; Tollefsen 2005; Tollefsen and Dale 2012). Clearly, the minimalistic account of joint action required here cannot appeal to propositional attitudes such as the intention to do *A*, the belief that *P*, linguistically represented collective goals, and so on. The section "A New Minimalism for Practice Theorists" will begin with a discussion of the sort of representations to which a minimalist account of joint action may appeal, if it is to ground Brandom's inferentialist-expressivism in a noncircular way.

Second, minimalist accounts typically try to account for joint action with resources that might be available either to prelinguistic children or to nonlinguistic animals. However, prelinguistic children and nonlinguistic animals are importantly different in the present context. For example, in her elaboration of a minimal "evolving collective intentionality," Satne says "This view takes as its point of departure the apparent fact that children are immersed in social practices from the beginning of their lives" (Satne 2021a, p. 110). This is a consequential starting point. Human children are already the product of an evolutionary process that has produced language and full-blown intentionality. Humans have evolved so that children are predisposed to notice the regularities of language, to have pro-social attitudes, and to exhibit normative attitudes. These inborn capacities are then scaffolded by the existing social practices and language. From the point of view of the social ontology objection, however, a minimalist account of joint action that appeals to the capacities of modern children just pushes the problem back one step. How is it that a basic linguistic practice of the sort analyzed by Brandom can emerge at all? Clearly, rebutting the social ontology objection requires an account of joint action invoking only capacities that are either compatible with the capacities we see in nonlinguistic social animals or are plausible evolutionary refinements of these capacities.[1]

"A New Minimalism for Practice Theorists" will develop an account of joint action that satisfies the two criteria just identified. It will argue that the proposed analysis includes those cases of joint action that clearly fall within the scope of the neutral description and excludes those that fall outside of it. The minimalist account thus stands on its own as an analysis of joint action. The section "A Social-Ontological Foundation for Practice Theory" will then apply the analysis to Brandom's practice theory. It will argue that supplementing Brandom's account with the minimalism not only rebuts the social ontology objection, the resulting social-ontological foundation sheds light on some of the darker corners of Brandom's account. Moreover, because the minimalist account of joint action invokes only capacities available to our pre-hominid ancestors, it puts Brandom's practice theory into a broadly naturalistic framework, helping us understand how linguistic and intentional beings like us could emerge from social, but nonlinguistic, animals.

A Neutral Description of Joint Action

The philosophical program of social ontology has proposed several analyses of what it means for agents to act together in a robust sense. Analyzing joint action requires a pre-theoretical identification of the phenomenon to be analyzed, since an analysis is adequate if and only if it includes paradigmatic cases of joint action, excludes paradigmatic individual actions, and adjudicates borderline cases in an illuminating way. In much

[1] Respecting this constraint should not be understood as discounting the importance of an evolutionary perspective on practices. The analysis of this essay is consistent with and complementary to Rouse's (2023) "natureculture" approach to practice theory.

philosophical analysis, the phenomenon is already characterized in ordinary language. But because "joint action" and "joint intention" are philosophical terms of art, there is no ordinary language meaning to fall back upon. As a result, identification of joint actions has been heavily influenced by philosophical theories of joint action. The canonical examples in the literature are simply those that all currently popular theories agree upon. For the sake of evaluating the force of the social ontology objection, this situation is unsatisfactory. In the absence of a neutral delimitation of the class of actions counting as "joint"—that is, a delimitation independent of any philosophical theory—it is impossible to tell whether practice theory is deeply flawed or merely inconsistent with some philosophical theories of action.

Up to this point, the phrases "joint action," "joint intention," "collective intention" and so on have not been distinguished. It is time to determine just what our object of analysis needs to be. As a characterization of the target phenomenon, "intention" (whether joint, collective, we-, or otherwise qualified) is unsatisfactory. "Intention" suggests a psychological state that might or might not be realized in action. From the point of view of practice theory, treating intentions as special facts about the mind is one of the vexing pictures we would like to overcome. Moreover, philosophers standardly distinguish actions from events by appeal to intentions. From the perspective of practice theory, to identify a special class of events as arising from intentions, and then ask for an analysis of intentions, is already too theory-laden to serve as a neutral description of the phenomenon to be analyzed. "Joint action" is a much thinner description than "joint intention," and we will therefore take it as our focus.

A theory-neutral description of joint action must be *superficial* in the sense that it does not presuppose the analytical apparatus of any particular philosophical account. Indeed, a description so superficial as to be nothing but platitudes would have substantial dialectical strength. To remain superficial, we will focus on characteristics of action descriptions. This strategy relies on the time-worn point that a single event can be intentional under one description and unintentional under another (cf. Davidson 1963). The phrase "joint action" suggests a pair of platitudes: under a description and in an interpretive context, joint actions are *joint* and they are *intentional*. This section will argue that these two platitudes can be unpacked in a way that does not presuppose one or another philosophical analysis of joint action. That is, we will argue that the two platitudes successfully circumscribe paradigm cases of joint action and exclude paradigm cases of individual action. And while we are aiming for clarity, we expect the class of joint actions to have fuzzy boundaries. If the boundaries were crystal clear, there would be little interest in an analysis that adjudicated borderline cases.

To say that an action is *joint* is to describe it as involving multiple agents. This suggests that the subject of the action description will typically be either a plural pronoun, for example, "We won the championship," or a noun that refers to more than one agent, for example, "The British army won the Battle of Waterloo." So far so good, but the plurality of the subject term does not isolate the class of joint actions. There are some action descriptions with plural subjects that characterize a multiplicity of individual actions, for example, "Susan and Nadia won the Twin Lakes Annual Tennis Championship." said of successive winners of a singles tournament. And there are uses where the subject of the sentence is singular, even though the action is joint, for example, "Wellington won the Battle of Waterloo."

By way of clarifying the contrast between individual and joint actions, Butterfill (2017, p. 361) helpfully invokes the distinction between the *distributive* and the *collective* reading of a predicate. A predicate, F is read distributively if and only if it is true of a number of individuals $\{a_1, \ldots, a_n\}$ and for each individual, a_i, Fa_i is true. The predicate in "The books are on the shelf" is naturally read distributively: it is true of each book that it is on the shelf. A predicate is collective when not distributive. In the sentence, "The legs support the tabletop," the predicate "...supports the tabletop" is not true of any leg individually, but only of the legs considered together. "Susan and Nadia won the Twin Lakes Annual Tennis Championship," similarly is distributive when said of multiple singles tennis competition winners. However, said of a doubles competition, where Susan and Nadia were partners, the predicate would need a collective interpretation. Some predicates, like "... won the Battle of Waterloo" demand collective interpretations, since it is impossible for an individual to win a military battle by themselves. When predicated of an individual, for example, "Wellington won the Battle of Waterloo," it is literally false but charitably read as a metonym.

The second platitude is that joint actions are intentional. Philosophical habit tempts us to suppose that intentional actions are those where the agent acted because of a belief, desire, plan, goal, or something similar. While uncontroversial (in most circles), such a characterization is not superficial enough. It invokes a particular analysis of what it means to act intentionally. Moreover, it privileges deliberate action by adults, leaving spontaneous, habitual, or otherwise unreflective actions—not to mention the actions of animals and prelinguistic children—in limbo. A better approach to the intentional/unintentional distinction is provided by Austin's "Plea for Excuses" (1961). He notes that, in common usage, "unintentional" does the heavy lifting of the intentional/unintentional pair. We have a rich practice for articulating excuses, and excuses show the variety of ways in which an action or its effects might be unintentional. We plead accident, mistake, bad luck, or unforeseeable consequence. And in the twenty-first century, we can add unconscious biases and other neural mechanisms to Austin's list. The domain of intentional action, Austin argues, is simply what is not unintentional.

Austin's insight can be married to Davidson's point that actions are intentional under a description. Action predicates like "…won the doubles tournament" or "…painted the house together" are correctly applied only in cases where the outcome was not a mistake or accident. Therefore, if an event is described with an action verb, then in the context, Austinean excuses have already been excluded. Austin's platitudinous characterization of intentionality thus dovetails with the emphasis here on action descriptions. Our two platitudes—that joint actions are both *joint* and *intentional*—thus lead to a plausible, theory-neutral characterization of the phenomenon of joint action:

> A joint action is one that is intentional under a description where (1) the subject of the descriptive sentence refers to multiple participants and (2) the predicate is an action verb interpreted as a collective predicate.

It is important to bear in mind that the satisfaction of clauses (1) and (2) cannot be read off the surface syntax or lexical semantics of the description. The interpretation may depend on knowledge of the context, since many sentences will describe individual actions in some contexts and joint actions in others, just as "...won the tennis championship" describes an individual action in a singles tournament, but a joint action in doubles.

Since the neutral characterization suggested above is not an analysis, it does not attempt to draw a bright line around the class of joint actions. There are many gray areas, and some examples will depend strongly on the way the details are described. This is to be expected, since if the phenomenon were already perfectly clear, there would be little need for an analysis. Nonetheless, a satisfactory description of the phenomenon to be analyzed ought to clearly capture the paradigm cases and provide some rough guidance for adjudicating borderline cases. To test the proposed neutral description, let us compare it to some paradigm cases.

In a recent essay on joint action, Butterfill (2017) opens with a rundown of paradigm cases from philosophical and psychological sources (citations omitted):

> There are phenomena, call them *joint actions*, paradigm cases of which are held to involve two people painting a house together, lifting a heavy sofa together, preparing a hollandaise sauce together, going to Chicago together, and walking together. In developmental psychology paradigm cases of joint action include two people tidying up the toys together, cooperatively pulling handles in sequence to make a dog puppet sing, bouncing a block on a large trampoline together, and pretending to row a boat together. Other paradigm cases from research in cognitive psychology include two people lifting a two-handled basket, putting a stick through a ring, and swinging their legs in phase. (Butterfill 2017, p. 357)

While two cases on this list need further discussion, the rest of Butterfill's examples fall squarely within the scope of our neutral description of joint action. The items drawn from the psychological literature deploy experimental protocols designed to elicit actions that are intentional under the indicated description. The protocols also guarantee that predicate will bear a collective interpretation, viz., that it cannot be true that "I made the dog puppet sing." In the examples from philosophical essays, the use of the word "together" in the description is intended to force the collective interpretation of the predicate. The two cases needing further discussion are "walking together" and "swinging their legs in phase."

When Margret Gilbert discusses walking together as a joint action (e.g., Gilbert 1990), she narratively constructs a substantial relationship between the parties. Indeed, their relationship is so strong that they are entitled to rebuke each other when, say, one party goes off in another direction without comment. Under the rich conditions described by Gilbert, the predicate "…walking together" is naturally read collectively. On the other hand, it is easy to construct examples—Gilbert herself does so—where "…walking together" has a distributive interpretation, and the example is not a joint action. Walking together, then, is a borderline case, and its status as a joint action depends heavily on the context. In this way it is unlike a clear paradigm case, such as lifting an object too heavy for one person.

The final example on Butterfill's list, two people swinging their legs in phase, is also at best a borderline case. With this example, Butterfill is gesturing at the growing body of research on coordination mechanisms. Both humans and other animals have neural mechanisms that permit the synchronization of action. This research needs to be treated with care. The essay from which Butterfill drew the example described the experiment in this way: "two seated participants were asked to visually coordinate their lower legs in either in phase or antiphase while oscillating them at a tempo of an auditory metronome pulse" (Schmidt and Richardson 2008, p. 284). Since the participants are trying to coordinate, it is tempting to interpret the action as joint. However, while "we are swinging our legs in phase to the pulse" is intentional, it will also be true that "I am swinging my legs in phase to the pulse". Hence, the predicate bears a distributive interpretation, even when the subjects are trying to coordinate. Moreover, another experiment described by Schmidt and Richardson (2008, p. 290) elicited the same mechanisms and behavioral result, and it was also not a joint action. In this case, the experimenters were looking for spontaneous coordination. Participants sat in adjacent rocking chairs and were *not* told to coordinate. The pairs ended up rocking together, but it is not true that they did so intentionally. A substantive conclusion thus follows from the neutral description: some forms of coordinated behavior produced by dynamic coordination mechanisms fail to count as joint actions.

A further useful feature of our neutral description is that it applies to both deliberate and spontaneous actions. The one-sided diet of examples in both the action theory and social ontology literature has tended to focus philosophical analyses on cases of deliberate action. Indeed, Bratman now characterizes his account as a "planning theory of acting together" Bratman (2013). But as Bratman and the others who contribute to these debates realize, some interesting kinds of joint action lack a plan or "prior intention." Having a description that is neutral between deliberate and spontaneous joint action is crucial for sharpening the social ontology objection. A practice theorist argues that the representational capacities necessary for planning arise out of shared activities. Therefore, on pain of an obvious circularity, the activities must be intentional, but not the result of deliberation or planning. Using the neutral criterion to identify joint actions, then, will identify the points where practice theory is vulnerable to the social ontology objection.

The neutral criterion of joint action circumscribes the phenomenon of joint action in a way that does not favor one analysis or another. It is entirely superficial, making no reference to underlying conditions that might explain the difference between joint and individual action. If an account of how representations emerge from social practices invokes actions that satisfy the neutral criterion, then the account is not merely inconsistent with a particular philosophical theory of joint action. It would be *prima facie* inconsistent with any account of joint action and thus vulnerable to the social ontology objection.

Inferentialism and Joint Action

Our question, now, is to determine whether and how practice theory depends on joint action. As argued in the Introduction, Brandom's version of practice theory is a useful case study because it is articulate about the character of the practices that ground representation. This section will argue that Brandom's account does require joint actions in the sense identified above, that is, it requires actions that are intentional under a description where the subject refers to multiple participants and the predicate must be interpreted as collective, not distributive. Moreover, it does so at a point where other versions of practice theory are also likely to make an appeal to joint action. Therefore, Brandom's practice theory is *prima facie* subject to the social ontology objection in a strong form. In the next section, we will develop a new, minimalist account of joint action, and argue that it provides Brandom a way out of the potential circularity.

Brandom's *Making It Explicit* (1994) is a systematic attempt to show how traditional representational vocabularies—including meaning, belief, intention, reference, and inference—can be understood in terms of their use in social practices. He sketches the approach this way:

> The explanatory strategy pursued here is to begin with an account of social practices, identify the particular structure they must exhibit in order to qualify as specifically *linguistic* practices, and then consider what different sorts of semantic contents those practices can confer on states, performances, and expressions caught up in them in suitable ways. The result is a new kind of conceptual-role semantics. (Brandom 1994, p. xiii)

The new kind of conceptual-role semantics is *inferentialist*. It understands meaning in terms of inference: the meaning of a proposition is the difference it makes to the conclusions one can draw from it and the premises from which it can be inferred. Since only propositions can serve as premises and conclusions, in the first instance, inferentialism provides meaning to propositions. The ingenuity of *Making it Explicit* is to render the meaning of sub-propositional units, such as predicates, referring expressions, and logical operators, in inferentialist terms as well. Such an account of meaning clearly cannot rely on a conception of inference that depends on logical rules defined in terms of logical operators, since doing so would be circular. Following Sellars, Brandom proposed that "material inference" is fundamental. The "particular structure" that a social practice "must exhibit in order qualify as a specifically linguistic practice," then, is that they support material inferences.

Brandom deploys the notions of commitment and entitlement to explicate the social practices that support material inferences. The content of a material inference can be understood as the commitments or entitlements it preserves. In the material analogues of deductive inferences, commitment to the premise requires commitment to the conclusion. In the material analogues of inductive inferences, commitment to the premise licenses entitlement to the conclusion. Commitments and entitlements are thus the building blocks of Brandom's account of content.

Commitment and entitlement are normative social statuses (Brandom 1994, p. 161). To call them social statuses is to say that commitment and entitlement are created when individuals take each other to be committed or entitled. Entitlements are constituted by the attribution of an authority to do something. Brandom is clear that an attribution of authority does not require language. It can be as simple as a practical partition of "the space of possible performances into those that have been authorized and those that have not, by being disposed to respond differently in the two cases" (Brandom 1994, pp. 161–2). The social status thus created is normative insofar as the different responses are permissions and sanctions. An individual *S* is entitled to do *A*, then, when others permit *S* to do *A*, while sanctioning other individuals who try to do *A*. For example, Jones is entitled to fish in the pond when Smith (and other members of the community) practically discriminate between those who are authorized and those who are not. Some who try to fish in the pond may be expelled while Jones is left alone.

While entitlements are constituted by attributions of authority, commitments are constituted by attributions of responsibility. Just as for entitlements, the content of a commitment is determined by the practical discriminations made by others. The structure of the practical discriminations, however, is more complex. First, "for someone to *undertake* a commitment ... is to do something that makes it appropriate to *attribute* the commitment to that individual" (Brandom 1994, p. 162). The first element of the practical discrimination, then, is that members of the community distinguish between those who have acted so as to undertake a commitment and those who have not. Second, taking on a responsibility to do *A* means that one makes oneself liable for criticism upon failure to do *A*. To undertake a commitment, then, is to entitle "those who attribute the commitment to sanction nonperformance" (Brandom 1994, p. 163). Such sanctions are made appropriate by the initial undertaking that licensed the attribution of commitment in the first place. The sanctions are thus normatively dependent on the undertaking. This means that the act of sanction is itself subject to practical discriminations of permission and punishment. Those who sanction nonperformance of *A* by an individual who is not committed to *A* are themselves subject to sanction. The structure of commitment, then, is that a person *S* undertakes a commitment to *A* when they do something that entitles others, S^* to sanction nonperformance. These others, S^*, are entitled to sanction nonperformance only if yet others S^{**} (perhaps including the person who undertook the commitment) take S^* to have the authority (be entitled) to sanction nonperformance of *A*. This in turn involves the sanctioning of S^* for inappropriate punishment. Since the person undertaking the commitment, *S*, must treat him/herself differently as a result of the commitment, this acknowledgment of the commitment is implicit insofar as *S* is among the S^{**} to entitle S^* to sanction nonperformance.

Returning to the social ontology objection, practice theorists, like Brandom, propose to explain our representational capacities in terms of some kind of social interaction, such as the social practice of language use. The objection complains that such social interactions constitute joint actions, and joint actions presuppose the very representational capacities that are to be explained by practice theory. Is Brandom's account vulnerable to this objection, and if so, at what point does the circularity arise? Brandom's expressivist inferentialism rests on a particular structure of social practices. The minimal social practices sufficient to get linguistic practices off the ground are webs of entitlements and commitments that support material inference: the practice of taking someone as committed or entitled to *Q* because they were committed or entitled to *P*. If Brandom's view is open to the social ontology objection, something in the performance of these social practices must require joint action.

Joint action was characterized above as one intentional under a description where the subject refers to multiple participants and the predicate is a collective predicate. The question now becomes whether Brandom's social practices are joint actions in this superficial sense. Entitlements and commitments are the basic elements, so we may start there. While entitlements and commitments are social statuses, not actions, they are the product of actions. Attributions, sanctions, permissions are all understood in practical terms, that is, as actions. The question, then, is whether entitlements and commitments can be the product of individual actions alone, or whether they require joint actions for their creation and maintenance.

To describe an action as attributing entitlement is to describe it as an intentional action, but not as a joint action. "S^* attributes to *S* an entitlement to *A*" refers to two individuals, but they are not both the subject of the action predicate; *S* is the object of the predicate and S^* is the lone subject. Similarly for descriptions of actions that constitute permissions (allowing *S* to *A*) and sanctions (punishing *S* for doing *A*). The actions that create and maintain the social status of an entitlement, therefore, are not joint actions and do not open Brandom's account to the social ontology objection.

Turning to commitments, recall the toy example of the "queen's shilling" that Brandom uses to introduce the idea:

> According to this practice, taking "the queen's shilling" from a recruiting officer counts as committing the recipient to military service. A performance of this kind has the same significance that signing a contract would have—in either case one has joined the army and undertaken all the commitments entailed by that change of status. (The official rationale was that some such overt irrevocable nonlinguistic performance was required to do duty for signing a contract, given that those enlisting were largely illiterate. The actual function of the practice was to enable "recruiting" by disguised officers, who frequented taverns and offered what was, unbeknownst to their victims, the queen's shilling, as a gesture of goodwill to those who had drunk up all of their own money. Those who accepted found out the significance of what they had done—the commitment they had undertaken, and so the alteration of their status—only upon awakening from the resulting stupor.) (Brandom 1994, p. 162)

By taking the queen's shilling, Jones joins the army. In Brandom's language, Jones has undertaken a commitment. It qualifies as a commitment because others (the recruiter, the other residents of the town, etc.)

attribute to Jones the responsibility to answer roll call. As with entitlements, describing Jones as undertaking a commitment (e.g., Jones took the queen's shilling) is an individual action because it has a single agent as its subject. Like bestowing an entitlement, *attributing* a commitment is not a joint action because its description does not have a plural subject. Therefore, attributing entitlements and undertaking commitments, along with their enabling actions, are all individual actions.

Plural subjects enter Brandom's story when webs of commitments and entitlements are structured into normative practices. Consider the material inference "Today is Tuesday, so tomorrow is Wednesday." For this to be a good inference, according to Brandom, there must be a practice wherein those who are entitled to utter "Today is Tuesday" are committed, by that utterance, to "Tomorrow is Wednesday." A community with such a practice may be said to endorse the inference "Today is Tuesday, so tomorrow is Wednesday." The action description "Speakers of English endorse the inference 'Today is Tuesday, so tomorrow is Wednesday'." has a plural subject, so it satisfies the first element of the neutral criterion for joint action.

The second element of the neutral criterion requires the predicate to be interpreted collectively, not distributively. Consider the predicate "…endorse the inference 'Today is Tuesday, so tomorrow is Wednesday'." While endorsing an inference is an action that an individual can do, Brandom explicitly denies that community assessment should be understood as a regularity of individual action within a group (Brandom 1994, p. 37ff). It is "norms all the way down," and there is no "reduction of normative status to nonnormatively specifiable dispositions, whether to perform or assess, whether individual or communal" (Brandom 1994, p. 44). To say that the inference "Today is Tuesday, so tomorrow is Wednesday" is endorsed by a community, then, is not to say that any member of the community does endorse it. Rather it is to say that within the community there is a structure of commitments and entitlements such that those who are entitled to assert "Today is Tuesday" are committed to "Tomorrow is Wednesday." As understood by Brandom, then, an action description like "Speakers of English endorse the inference 'Today is Tuesday, so tomorrow is Wednesday'." deploys a predicate that demands a collective interpretation. To describe someone as participating in a normative practice is therefore to describe them as participating in a joint action.

In spite of its sophisticated account of the social statuses underlying practical norms—a level of detail that will be useful below—Brandom's practice-theoretic account is vulnerable to the social ontology objection at just the point where other practice theories are vulnerable. Practice theorists typically gesture at community acceptance or assessment, which are clearly joint actions of some sort. Our deep dive into the details of Brandom's account shows that while much of the burden of instituting and maintaining normative practices is carried by individual action, a transition to joint action is necessary to institute linguistic practices.

A New Minimalism for Practice Theorists

Brandom's normative practice theory requires joint action. And since it invokes joint action at a point *prior* to the emergence of linguistic or mental representation, on pain of circularity, normative practice theory cannot depend on any analysis of joint action that invokes such representational states. The social ontology objection assumes that any account of joint action would invoke representational states, and it concludes that practice theory is circular. However, if there were a viable *minimalist* account of joint action, one that did not require sophisticated mental representations such as common knowledge or we-intentions with propositional content, the circularity could be avoided. This section will draw on the existing minimalist literature to provide such an analysis and thereby put practice theory on a firm foundation.

As argued in the Introduction, a minimalist analysis consistent with Brandom's view may invoke only representational capacities possessed by nonlinguistic creatures. As Brandom puts the point, such creatures are *sentient* but not *sapient*. An objection arises immediately: one might doubt that merely sentient animals *act* at all. Davidson (1982), for example, notoriously argued that beliefs require language, and without beliefs, animals could not form intentions. One might think, then, that the minimalist story is over before it begins. However, to argue that animals do not act intentionally on Davidson's kind of ground assumes a particular analysis of intentionality. The neutral criterion of joint action blocks this argument. In a neutral sense of intentional action, higher animals are appropriately described with action verbs: they *search* for food and mates, *hide* from predators. Such actions are distinct from events like a chameleon's change of color, or a lemming falling off a cliff. There is, then, a plausible distinction between intentional and unintentional action applicable to sentient, but non-sapient beings that does not depend on an analysis of intentionality like Davidson's.[2]

The foregoing response gives rise to a further challenge: if merely sentient animals are to be treated as acting intentionally, there must be some sense in which they represent their environment. But the goal of practice theory, one might argue, is to account for representation. Another circularity looms, or so one might think. However, such an objection saddles the practice theorist with an unfair explanatory burden. The target of practice theory is linguistic representation and the higher sorts of intentionality and thought that language makes possible. It does not preclude admission that sentient animals have the capacity to respond to their environment in ways that let them find food and water, identify mates, avoid predators, rear their young, and build shelters. This means that their brains must track changes in their environment, recognize affordances, and act for the sake of goals. The sentient, but non-sapient beings who populate Brandom's thought experiments in *Making It Explicit* also have the capacity to respond to actions of such conspecifics in ways that constitute permissions and punishments. Such capacities are well within the behavioral repertoire of social animals (Risjord 2006). Since we are focusing on the capacity for action, we need not get bogged down in the specific cognitive processes that underpin a sentient, non-sapient animal's capacity to act. Our benchmark must be the actions that sentient, non-sapient animals can demonstrably perform, and insofar as nonlinguistic animals must in some sense "represent" the affordances of their environment and their goals, any such "representations" will be consistent with a practice theoretic approach to representation.

---

[2] Hutto and Satne (2015) make a similar point in terms of "ur-intentionality." The difference between their view and the view developed here is that, for the reasons articulated above, the current problem demands a focus on action, rather than intention.

The minimalist analysis of joint action to be provided here, then, is bound by two constraints: it must postulate capacities available to non-linguistic (sentient, non-sapient) social animals, and it must provide necessary and sufficient conditions for actions that are joint under the neutral criterion. The discussion below will assume that more than one animal is involved in the action, so the dispositive question will be: what sort of interaction is required to motivate a collective reading of the action predicate, and when is a collective reading demanded? To make the discussion concrete, it will be useful to have before us an example of animal behavior that is ambiguous between joint and individualist interpretations. The ambiguity will let us identify those considerations that would force a collective reading or permit a distributive one. Chimpanzee hunting has been subject to much empirical investigation, and the debate has focused exactly on the question before us: when chimpanzees hunt, is it a joint action, or does each member of the troop act individually?

Based on his extensive ethological observations of chimpanzee hunting, Christophe Boesch distinguished the hunting strategies of two populations of chimpanzees, one in the Gombe Stream National Park, Tanzania and the other in the Taï National Park in Côte d'Ivorie (Boesch 1994, 2002). Both populations hunt red colobus monkeys, but they deploy different strategies.

> Whenever Gombe chimpanzees hunt in groups, they tend to hunt independently, each chasing a different target, usually in different directions. They could be described as performing 'simultaneous solitary hunts'.... (Boesch 1994, 1143)

By contrast, the Taï chimpanzees have a rather elaborate system of roles. When a group of chimpanzees from the Taï population encounters a group of colobus monkeys,

> ...one of [the chimpanzees] slowly climbs to a height of about 5 m, usually unnoticed by the colobus. ... The others move on the ground in anticipation of the possible escape routes of the colobus and ready themselves to join the pursuit. When the climber has been spotted by the monkeys, he rushes up, and they begin to move. The hunter's contribution then consists

> mainly of keeping them moving in a certain direction while the others follow on the ground and undertake different blocking moves. The driver usually follows the monkeys in the branches without trying to capture one on his own. (Boesch 2002, 32–33)

The *driver* role is thus distinct from the *blocker*, who places "himself in a tree so as to block the progression of the prey" (Boesch 2002, 33). The driver and blockers are joined by a *chaser*, who moves "quickly after the prey, trying his best to catch up with one" (Boesch 2002, 33). In many cases, these three roles are supplemented by an *ambusher*:

> The ambusher is the hunter that anticipates the escape route of the quarry long enough in advance to be able to force it to turn backwards towards its pursuers or to move it downwards into the lower canopy, where chimpanzees have a very good chance of catching it, because in the continuous tree cover at this level chimpanzees move faster than colobus monkeys. Thus an *ambusher* is a hunter placing himself in position where no prey is yet, and where he cannot be easily seen, and he will rush toward the prey as soon as it enters his tree. (Boesch 2002, 33)

On the neutral criterion of joint action, the Gombe chimpanzees are undertaking individual actions when they hunt, but the Taï chimpanzees are engaged in a joint action. To say that the Gombe chimpanzees are engaged in "simultaneous solitary hunts" is to give the action verb "…are hunting" a distributive reading: it is true of each chimpanzee that they are hunting, and many are hunting simultaneously. However, the differentiation of roles among Taï chimpanzees means that it is not true that each chimpanzee is hunting the monkey. The driver and blockers are not trying to capture the monkey. Insofar as it is true that the driver is hunting, they are hunting as a part of the group, not individually. Hence, the predicate "…are hunting" requires a collective reading for the Taï chimpanzees.

The difference between the Taï and Gombe chimpanzees is thus an excellent test case for a minimalist analysis of joint action. What capacities are being exhibited by the Taï chimpanzees that are not exhibited by the Gombe chimpanzees?

The first difference between the cases is that only the Taï chimpanzee hunting has a collective or joint outcome. In the Taï case, the capture of the monkey results from the chimpanzees acting together. By contrast, in the Gombe case, each individual is chasing a different monkey and an individual's success is the result of their own actions alone. Pacherie calls this difference the "common outcome requirement" for joint action. She characterizes the requirement in these terms: "an event qualifies as a joint action only if it is describable as the common outcome of what several agents did; that is, none of the agents involved did on their own bring about this outcome" (Pacherie 2013, 1819). Having a common outcome is a plausible necessary condition for the predicate "...are hunting" to bear a collective reading. But it is not yet sufficient. A group of Gombe chimpanzees might isolate a single monkey, and all pursue it. Since all of the chimpanzees are influencing the response of the monkey, its final capture would not be the result of a single chimpanzee acting on its own. Further conditions are required by a minimalist analysis.

Most minimalist analyses postulate further that the agents must represent the common outcome as a goal that they will achieve. Vesper et al. (2010) characterize this in a way that is arguably compatible with the assumption that the agents are sentient but not sapient. However, in the present context, even such a modest requirement is dubious. The joint actions in question for practice theory are the participation in a normative practice. But when participants engage in a normative practice, even participants who are fully sapient, it is not true to say that they have the goal of maintaining the practice, or of achieving the benefits that such a practice might confer on the community. There are two difficulties. First, in many practices with a common outcome, the participants have a variety of goals. Second, requiring that every joint action have a known, shared goal treats all joint actions as instrumental and deliberate. Since practices are not followed deliberately, requiring a shared goal would result in an uncomfortably narrow understanding of joint action. At the same time, it is clear that a collective outcome alone could not distinguish between an outcome achieved jointly, and an outcome achieved individually by each member of the group. The challenge is to make the distinction without postulating a collective goal.

To distinguish between collective outcomes achieved jointly and those achieved individually, the minimalist analysis proposed here requires that the collective outcome result from two processes, both of which force a collective interpretation of the action predicate: the collective outcome is the result of *joint attention* and *mutual coordination*. Neither process, we will argue, invokes representational states that would invite the circularity of the social ontology objection.

Joint attention is commonly discussed as a component of joint action for rather obvious reasons. Any action requires perceptual engagement with objects (including perhaps other agents) as either means or ends of the action. Were multiple agents each attending independently to the environmental affordances, the predicates of any description would have natural distributive readings. However, there has been much dispute over how to understand "joint attention" since the idea was introduced into psychology in the 1980s. Weaker readings, such as Leavens and Racine (2009) understand joint attention as "the intentional co-orientation of two or more organisms to the same locus," where "intentional co-orientation" means that at least one of the organisms does something to direct their gaze to the common locus (Leavens and Racine 2009, p. 241). Others, such as Tomasello (1995), require not only that the agents intentionally attend to the same object, they are each aware that the other is attentive. Joint attention, on such a view, rises to the level of mutual knowledge. From a minimalist perspective, the more permissive reading is preferable. As Leavens and Racine argue, there is evidence that the great apes engage in joint attention in the broad sense. Hence, the more generous conception of joint attention applies to animals that are sentient, but not sapient.

Advocates of a strong reading of joint attention point out that the permissive reading does not require that the agents are attending together in a way that is "truly joint" (Carpenter and Liebal, 2011, p. 159). It is true that joint attention, understood permissively, does not force a distributive reading on the action predicate "...are watching the crow fly." However, a minimalist account of joint action can accept this result. Joint attention is merely a necessary condition for joint action. Actions do not become "truly joint" unless the joint outcome is brought about by both joint attention and *mutual coordination.*

Both Pacherie's (2013) minimalist account of joint action and Bratman's (2013) more demanding account appeal to coordination among the agents. As used here, "mutual coordination" does *not* require the kind of inter-related intentions postulated by Bratman's analysis. Like "joint attention," "mutual coordination" is read broadly so as to include sentient, non-sapient agents. Mutual coordination has at least two dimensions. First, achieving the collective outcome requires the agents to take complementary roles, and they must recognize both their own role and its relationship to the other(s). Second, since each of the roles will have its own requirements for action, each of the participants must not only recognize the affordances for their own role, they must recognize the role-relevant affordances for the other(s) (Risjord 2016). And in many cases, these different affordances must be the subject of joint attention (think of the different roles of players in a fast break in basketball). Mutual coordination is necessary for a collective interpretation of the action predicate. Without some kind of coordination, the actions could be independent and the outcome independently achieved, and therefore the action predicate would bear a distributive reading.

We have argued that a collective outcome, joint attention, and mutual coordination are necessary for joint action. Where the outcome is the result of actions that involve joint attention and mutual coordination, these three conditions are also sufficient. To see why, suppose that these three conditions were not jointly sufficient. This would mean that the conditions were satisfied, yet the action was intentional under a description *A*, the description had a plural subject (say, s_1 and s_2), and that *A* is a *distributive* action predicate true of subjects s_1 and s_2. It follows from the distributive reading of *A* that neither "s_1 is *A*-ing" nor "s_2 is *A*-ing" is a joint action. However, if the minimal conditions for joint action were satisfied, then s_1 and s_2 *A*-ing involved joint attention and mutual coordination, and that the collective outcome was a product of joint attention and mutual coordination. But in such a case, neither s_1 nor s_2 could have *A*'d all alone. It follows that the action predicate must have a collective reading, thus contradicting our original supposition. Bringing about a collective outcome through activity that requires both joint attention and mutual coordination is therefore sufficient for joint action.

The minimalist account of action permits us to understand the Taï chimpanzees as acting jointly, while the Gombe chimpanzees are not. The Taï chimpanzees not only achieve a joint outcome that no one chimp does alone, according to Boesch's account, they do so through joint attention and mutual coordination. A strength of the foregoing account is that it permits a response to the "social tool" objection often raised against *prima facie* examples of animal joint action. This objection suggests that just as the chimpanzees can use inanimate objects as tools, an individual chimpanzee can recognize that a conspecific might have an instrumental use, that is, by blocking the retreat of a monkey. The Taï chimpanzees are thus treating each other as "mere" social tools, not as genuine agents with whom they are participating in an action. In response to this objection, the minimalist analysis provides the resources for distinguishing cases where the animals are using each other instrumentally from those where there is genuine joint action. Where one animal uses another merely instrumentally, then joint attention and mutual coordination based on joint attention will be missing. However, if Boesch's description is correct, the Taï chimpanzees are engaged in (minimal) joint action.

The minimalist analysis of joint action proposed here is consistent with the more cognitively demanding analyses proposed by non-minimalist social ontologists. These more demanding analyses of joint action can be understood as applying to narrower domains of joint action, such as those involving planning, negotiation, deliberation, or a rich institutional context. For sapient beings like us, who mature into a world where language and social structures already exist, many of our joint actions have this richer, representationally inflected character. Nonetheless, we retain the capacity for spontaneous joint action, and at least some of our social coordination relies on the same minimal mechanisms deployed by social animals and young children (Risjord 2014). From the perspective of practice theory then, we should see the richer kind of joint action made possible by language and full-blown sapience as emerging out of and informing the minimal capacities that make the practice possible at all.

A Social-Ontological Foundation for Practice Theory

At the end of "Inferentialism and Joint Action," we noted that Brandom's version of practice theory depends on joint action at the same point as other practice theorists: at the level of community acceptance or endorsement. However, while practice theorists like Sellars and Wittgenstein leave community endorsement of norms rather opaque, Brandom fleshes out the interactions that underpin normative practices. This level of detail permits us both to show how the minimalist analysis of joint action can block the social ontology objection and shed light on what "community endorsement" amounts to.

For Brandom's inferentialism, a community endorses a material inference when there is a web of commitments and entitlements in place. As shown above, undertaking commitments and attributing entitlements are individual, not joint, actions. But maintaining the web of commitments and entitlements that constitute a normative practice is not something that one person can do. Maintaining the practice is thus a collective outcome. According to the minimalist analysis presented above, if this is to be a joint action, it must be brought about by a process involving joint attention and mutual coordination.

Brandom's normative practices could not emerge from the attribution of entitlements and undertaking of commitments without both joint attention and mutual coordination. If agent S^* attributes an entitlement to agent S to do action A, then S^* must *permit* S to do A and *sanction* those not entitled. Permissions and sanctions are normative, and that means that S^* is treated as entitled to permit or sanction by yet other agents S^{**}. Permissions and sanctions would be impossible unless S^* and S^{**} could jointly attend to S's behavior. Without joint attention, there would be no sense in which S^{**} was permitting/sanctioning S^*'s permission/sanction of S. Moreover, S^*'s attribution of an entitlement to S and S^{**}'s attribution of a corresponding entitlement to S^* are two different roles necessary for the actions that maintain S's entitlement. These roles must be coordinated in such a way that S^*'s actions result in an entitlement for agent S, rather than someone else. Commitments, too, depend on permissions and entitlements, and thereby are the outcome of mutual coordination and joint attention.

The minimalist account of joint action thus provides a social-ontological foundation for Brandom's practice theory. Maintaining the web of social statuses that constitute a normative social practice requires joint actions, as the social ontology objection notes. But the circularity attributed to practice theory by the social ontology object does not arise because these joint actions can be analyzed by the minimalist account. The minimalist account does not appeal to intentions or other linguistically inflected representations that would render Brandom's normative practice theory circular. The social ontology objection, therefore, does not apply to Brandom's version of practice theory. (Whether it applies to Sellars or Wittgenstein, of course, is an open question, but there are some grounds for optimism.)

In addition, the minimalist foundation arguably demystifies what it means to be a participant in a normative practice. Participants need to jointly attend to the actions of others and take the roles that institute permissions and sanctions, commitments, and entitlements. Insofar as an individual is subject to the normative attitudes and exhibits them toward others, they are a member of the community that supports the normative practices. The community is thus a network where each individual is a node, and the actions attributing entitlements and undertaking commitments are the relations. And as we saw above, each action of attributing an entitlement or undertaking a commitment is an individual action. The jointness arises from the relationships among the individual actions constituted by joint attention and mutual coordination. There is, then, a quite tractable sense in which a community can endorse a norm without any member forming intentions with propositional content.

Conclusion

In a recent discussion of how Sellars's practice theory relates to his account of collective intentionality, Peter Olen and Stephen Turner explore Sellars's relationship to Durkheim's collective mind. They argue that Sellars is caught between his empiricist scruples and his need for something trans-individual to ground his account of "ought." They conclude that:

> The language of commitment, decision, and acceptance points to the difficulty with [Sellars's] conception of the problem. If we were to grant a notion of a group mind, we could allow it to make collective decisions, such as a decision to accept a given symbolic system for the collective 'us'. Sellars excludes this possibility on the reasonable grounds that it conflicts with basic empiricist commitments in the philosophy of mind. But this leaves him only with *individual* acts of acceptance, decision, and commitment. *And these individual acts cannot produce norms that bind others: they are not collective.* (Olen and Turner 2015, 973, emphasis in original)

While we have not been concerned with the *normativity* of practices, this essay shows that there is a possible third way out of Olen and Turner's conundrum. Attributing entitlements and undertaking commitments are indeed actions done by individuals. Bringing about joint outcomes through joint attention and mutual coordination produces joint actions that require no group mind or other trans-individual intentional capacities. The capacities required for such minimal joint actions are entirely consistent with empiricist—or better, naturalist—commitments in the philosophy of mind. Moreover, because it is possible, at least in principle, for nonlinguistic animals to exhibit minimal joint action, the social-ontological foundation for practice theory also provides the basis for an evolutionary account. If Brandom's normative pragmatics is sufficient for the creation and maintenance of norms, then we have, in principle, an evolutionary pathway through which human normativity could have evolved.

References

Austin, J.L. 1961. *Philosophical Papers*. Clarendon Press.

Boesch, C. 1994. Chimpanzees-Red Colobus Monkeys: a Predator-Prey system. *Animal Behaviour* 47 (5): 1135–1148.

———. 2002. Cooperative Hunting Roles Among Taï chimpanzees. *Human Nature* 13 (1): 27–46.

Brandom, R. 1994. *Making it Explicit: Reasoning, Representing, and Discursive Commitment*. Harvard University Press.

———. 2008. *Between Saying and Doing: Towards an Analytic Pragmatism*. Oxford University Press.

———. 2009. *Articulating Reasons*. Harvard University Press.

Bratman, M.E. 2013. *Shared agency: A Planning Theory of Acting Together*. Oxford University Press.

Butterfill, S. 2011. Joint Action and Development. *The Philosophical Quarterly* 62 (246): 23–47.

———. 2017. Joint Action: A Minimalist Approach. In *Routledge Handbook on the Social Mind*, ed. J. Kiverstein, 357–369. Routledge.

Carpenter, M., and K. Liebal. 2011. Joint Attention, Communication, and Knowing Together in Infancy. In *Joint Attention: New Developments in Psychology, Philosophy of Mind, and Social Neuroscience*, ed. A. Seeman, 159–181. MIT Press.

Davidson, D. 1963. Actions, Reasons, and Causes. *Journal of Philosophy* 60 (23): 685–700.

———. 1982. Rational Animals. *Dialectica* 36 (4): 317–327.

Fodor, J., and E. Lepore. 2007. Brandom Beleaguered. *Philosophy and Phenomenological Research* 74 (3): 677–691.

Gilbert, M. 1990. Walking Together: A Paradigmatic Social Phenomenon. *MidWest Studies in Philosophy* 15 (1): 1–14.

Harman, G.H. 1968. Three Levels of Meaning. *The Journal of Philosophy* 65 (19): 590–602.

Hutto, D., and G. Satne. 2015. The Natural Origins of Content. *Philosophia* 43 (3): 521–536.

Koons, J.R. 2021. Sellars on Rational Agency as Presupposing Collective Attitudes. In *Groups, Norms and Practices: Essays on Inferentialism and Collective Intentionality*, ed. L. Koreň et al., 189–213. Springer Nature.

Koreň, L., et al., eds. 2021. *Groups, Norms and Practices: Essays on Inferentialism and Collective Intentionality*. Springer Nature.

Leavens, D., and T.P. Racine. 2009. Joint Attention in Apes and Humans: Are Humans Unique? *Journal of Consciousness Studies* 16 (6–8): 240–267.

Loeffler, R. 2009. Neo-Pragmatist (Practice-Based) Theories of Meaning. *Philosophy Compass* 4 (1): 197–218.

———. 2021. Implicit Scorekeeping: A We-Mode Account of Belief and Interpretation. In *Groups, Norms and Practices: Essays on Inferentialism and Collective Intentionality*, ed. J.R. Koreň et al., 59–81. Springer Nature.

Olen, P., and S. Turner. 2015. Durkheim, Sellars, and the Origins of Collective Intentionality. *British Journal for the History of Philosophy* 23 (5): 954–975.

Pacherie, E. 2013. Intentional Joint Agency: Shared Intention Lite. *Synthese* 190 (10): 1817–1839.

Risjord, M. 2006. Evolution and the Kantian Worldview. *The Southern Journal of Philosophy* 44 (S1): 72–84.

———. 2014. Structure, Agency, and Improvisation. In *Rethinking the Individualism/Holism Debate: Essays in Philosophy of Social Science*, ed. J. Zahle and C. Finn, 219–236. Springer Synthese Library.

———. 2016. Ecological Attunement and the Normativity of Practice. In *Normativity and Naturalism in the Philosophy of the Social Sciences*, ed. M. Risjord, 175–193. Routledge.

Rouse, J. 2023. *Social Practices as Biological Niche Construction*. University of Chicago Press.

Satne, G. 2016. A Two-Step Theory of the Evolution of Human Thinking. *Journal of Social Ontology* 2 (1): 105–116.

———. 2021a. Understanding Others by Doing Things Together: An Enactive Account. *Synthese* 198 (1): 507–528.

———. 2021b. Collective Intentionality, Inferentialism and the Capacity for Claim-Making. In *Groups, Norms and Practices: Essays on Inferentialism and Collective Intentionality*, ed. L. Koreň et al., 99–118. Springer Nature.

Schmidt, R.C., and M.J. Richardson. 2008. Dynamics of Interpersonal Coordination. In *Coordination: Neural, Behavioral and Social Dynamics*, ed. A. Fuchs and V.K. Jirsa, 281–308. Springer.

Tollefsen, D. 2005. Let's Pretend! Children and Joint Action. *Philosophy of the Social Sciences* 35 (1): 75–97.

Tollefsen, D., and R. Dale. 2012. Naturalizing Joint Action: A Process-Based Approach. *Philosophical Psychology* 25 (3): 385–407.

Tomasello, M. 1995. Joint Attention as Social Cognition. In *Joint Attention: Its Origins and Role in Development*, ed. C. Moore and P.J. Dunham, 103–130. Hillsdale, NJ: Lawrence Erlbaum.

Vesper, C., S. Butterfill, G. Knoblich, and N. Sebanz. 2010. A Minimal Architecture for Joint Action. *Neural Networks* 23 (8–9): 998–1003.

"Practices, Normativity, and the Natural History of Human Biological Niche Construction"

Joseph Rouse

Practice-based social theories have been greatly influenced by Wittgenstein's discussions of rule-following. Many practice theorists take Wittgenstein to have indicated a normative ordering and responsiveness expressed *in* what people do that is more fundamental than explicit interpretation *of* what they do. Participation in practices provides the background orientation, understanding, or competence that enables following rules, answering to norms, and articulating and grasping meanings. These capacities are concretely embedded in interrelations among a practice's performances or enactments. Social practice theories have been less attentive to Wittgenstein's suggestion that he is offering "remarks on the natural history of human beings" (1953, para. 415; also para. 25). Recent developments in evolutionary biology nevertheless provide a richer context for expanding on the latter remark, albeit in a stronger sense than

J. Rouse (✉)
Wesleyan University, Middletown, CT, USA
e-mail: jrouse@wesleyan.edu

K. M. Cahill (ed.), *Wittgenstein on Practice*,
https://doi.org/10.1007/978-3-031-68656-6_12

Wittgenstein likely intended.[1] My topic today concerns the implications of this re-conception of social practices for understanding normative accountability. The diversity of normative concerns arising in people's practice-differentiated ways of life is an evolved, "two-dimensional" form of the one-dimensional biological normativity of most organismic lineages. This recognition has important implications for how normative concerns arise and interact, *how* their content is embedded in practices, and how those concerns are expressed discursively.

I develop these themes in four parts. The first part provides the background for my account drawn from recent developments in evolutionary biology and the social theory of practices. The second part then identifies a constitutive ambiguity in the normativity of practices, between the sense of what participants normally do that makes their performances predictable for other participants, and an authoritative sense of what they ought to do. This part of the chapter argues that these are not *alternative* senses of normativity, however, but instead must work together to constitute how practices sustain and revise the normative ordering of human ways of life. The third part of the chapter builds on the dynamic interaction of these two senses of normativity to show why the normativity of practices is temporally extended, as accountability to what is at issue and at stake in the future development of prior patterns of situated performance rather than to already-determinate norms. The final part explicates the characteristic "two-dimensional" normativity of our practice-differentiated way of life, which had been introduced in the opening stage-setting, but can now be developed in light of the two intervening parts of the chapter.

Stage-Setting

This integration of social and biological conceptions of practices and their normativity draws on an ecological-developmental conception of organisms (Gilbert 2016; Sultan 2015). Organisms are not self-contained

[1] I develop this conception of practices and their normativity as biological phenomena in detail in Rouse (2023).

individual bodies, but processes that incorporate their developmental and selective environments. An organism's environment is not its physical surroundings, but only those features involved in sustaining and reproducing its constitutive processes. Social organisms inhabit integrated social-ecological environments. Their conspecifics—not just their presence, but their postures, activities, and responsiveness—are consequential aspects of those surroundings, and often closely entangled with other circumstances.

Evolutionary biology has long emphasized how environments exert selection pressures that transform organismic lineages. Biologists now recognize organism-environment relations as a two-way process. Niche construction is the process through which organisms act on their environments and change the selection pressures on their own and other lineages (Odling-Smee et al. 2003). It includes behavioral niche construction—forms of behavior that generate selection pressures to produce descendants of that behavior in subsequent generations. Human languages are probably the pre-eminent example of behavioral niche construction.

"Classical" practice theories (e.g., Bourdieu 1990; Giddens 1984) identified practices as constituted by performative regularities, shared presuppositions, or governing norms. Recent accounts more commonly abandon general characterizations of practices in favor of particularist examples (Turner 2007). My account differs from both approaches. What identifies and sustains a practice is not any underlying commonality. Practices are instead marked by the closely coupled interdependence of their situated performances. Each performance is oriented by how it aligns with others in mutually responsive and often supportive ways in appropriate circumstances. These patterns originated in the evolutionary novelty of a cooperative way of life, in lieu of the merely coordinated behaviors of other social primates (Sterelny 2012).

Humans now thereby live a practice-*differentiated* way of life. We do not merely depend on others' performances in shared practices. Those practices in turn depend on effective alignment with other practices at more limited interfaces. Most of our biological needs are now met by practices in which we do not participate, and the practices in which we do participate depend on other practices for many other forms of support. This extended nexus of practices is the basic shape of human developmental environments, differentiated along two axes. Human ways of life are horizontally differentiated. We each live very different lives according to which practices we take up. Each person's life is also vertically differentiated by participation in multiple practices which they must accommodate together. The ongoing development of practices is often shaped in part by interactions between these two forms of differentiation.

In its broadest sense, normativity concerns what is nonarbitrarily open to assessment, whether as good or bad, successful or failing, correct or incorrect, meaningful or senseless, just or unjust, and so forth. Organisms are only open to such assessment in a limited way. They are Aristotelian *energeia*, goal-directed processes whose goal is the continuation of that very process. The physiology, development, behavior, and evolution of most organisms are thereby only nonarbitrarily normative along a single dimension of success or failure in sustaining life and producing living descendants. Biological lineages that succeed in reproducing themselves do so with the same biological import, as successfully adapted to changing environments whether they do so by remaining relatively stable in their morphology, development, and behavior over time, or by changing dramatically in one or more respects in response to changes in selection pressures. Evolution is similarly indifferent to whether lineages sustain themselves as "generalists" that can subsist in varied environments or become tightly adapted to specific environmental conditions. Evolutionary success consists in producing descendants in subsequent generations, period. What matters biologically is the successful continuation of *this* life and *this* lineage, in *whatever* forms they take subsequently.

The evolution of a practice-differentiated way of life introduces a second dimension of nonarbitrary normative assessment into life cycles of the hominin lineage. Like organisms in this respect, practices are patterns of environmental intra-action which only continue to exist through their ongoing reproduction. That reproduction is also goal-directed in a similar way: the goals of ongoing practices are internal to their continuing reproduction. The closely coupled interdependence of people's situated performances which sustains a practice thereby enables us to undertake activities, develop skills, and achieve outcomes not possible without that mutual support in conducive circumstances. The "goods" thereby achieved, that is, those practices' constitutive ends, are often only recognizable as such by those whose developed skills and discernments make those goods salient and intelligible. A new dimension of normative concern initially arises within practices as having their own nonarbitrary ends. Practices are nevertheless also components of a practice-differentiated way of life, and so they are also open to assessment for their contributions to other practices and their encompassing way of life. That opens a second dimension of normativity, rather than just a second component, because survival of the lineage no longer stands alone, but is now conjoined with other normative concerns arising from the mutual interdependence of horizontally differentiated practices.

The Constitutive Ambiguity of Normative Concerns

Practices are temporally extended patterns of situated, interdependent performance. What each participant does depends not only on past performances, but also on anticipated future continuation of the practice and other practices on which it depends. Practices must have sufficient prior stability to establish a projectible pattern, but that initial stability only enables further continuation of a practice. It cannot determine whether or how the practice continues, as Wittgenstein rightly insists (1953, para. 201–202). The normative "force" binding participants to a practice, however, is that internal, normative goal-directedness: people participate in practices for the sake of what the practice is doing, what it can or will do, and how their participation affects or shapes their own developing lives.

The need to sustain practices through multiple generations of participants highlights a constitutive ambiguity in the normativity of practices. Current participants initiate prospective new ones into the practice's activities, skills, and normative concerns, whether by instruction, ostension, or inadvertent modeling. To recognize a practice as responsive to a normative concern, in one sense, is for that concern to play a predictive role in how the practice continues. Because practices are normatively oriented, one can predict what participants will *normally* do in various circumstances, by predicting they mostly will do whatever is "called for" by that concern. In another sense, however, normative concerns always outrun or exceed this predictive sense of what is normal. What it is for a practice to go on "in the same way" as its prior performances is not definitively settled by those performances alone. The question of how that concern applies to subsequent cases is always at issue in a practice's subsequent development (Brandom 1979). In that sense, answering to normative concerns always requires authoritative exercise of judgment against the background of past performances.

These two aspects of normativity roughly correspond to what Charles Taylor (1995, 167–68) once described as "two broad schools of interpretation of Wittgenstein" on how obeying a rule is a practice (Wittgenstein 1953, para. 202). Sustaining and reproducing practices over time nevertheless requires that *both* the predictive and the authoritative senses of normativity work together, so that neither reading of Wittgenstein is sufficient by itself. New participants acquire the requisite postures and skills to take up a practice by learning to recognize practice-relevant differences in circumstances and what participants then ("predictively") ought to do. Such recognition can be initially guided by others, but new participants must eventually recognize such matters on their own. Without sufficient predictive reliability to enable (some) new participants to pick up on the practice's concerns and develop the requisite postured orientations for themselves, perhaps with guidance and correction from others, those practices would not be sustainable. The predictive normativity of practices also helps sustain the interdependence of its different performances and roles and their more limited dependence on other practices. Participants normally depend on what others do, and they must therefore rely on others to act in sufficiently predictable ways. Similarly, if performances of one practice depend on what is accomplished or determined by other practices, then the latter must have sufficient stability and predictive reliability for participants in the former to anticipate and rely on *mostly* proper performance of the latter.

While the predictive sense of normativity thus plays a necessary role in sustaining practices, it is not sufficient to understand how they are held together normatively. One reason for its insufficiency is that, although what participants do is partially dependent on their predictive grasp of how other participants will respond to a situation and how other practices will be carried out, their subsequent performances are responsive to what then happens, and to how their predictive sense of its normative concerns changes accordingly. Similarly, people participate in practices with an anticipatory orientation toward their predictive sense of how other practices will affect what they do. They nevertheless also then *normally* defer to the judgment or performance of that practice's participants if it conflicts with their anticipation. The requisite predictive character of a practice may therefore change over time while still maintaining its continuity as a practice sustained by the interdependence of its situated performances.

The predictive, normalizing sense of normativity is thus not the primary sense in which normative concerns are constitutive for practices, even though it plays important roles for initiates and outsiders. Someone who takes up a practice as a novice does so "successfully" only in the transition from merely imitating experienced participants to recognizing for themselves which considerations are relevant and how to respond, and acting accordingly. To participate in a practice is thus to acknowledge its normative concerns as authoritative *in practice* for one's own performances. To be a nurse, for example, to *participate* in the practice rather than merely imitate its performances, is not to attend to and care for the well-being of patients because that is what nurses do. Rather, nurses do what is needed to attend to and care for the well-being of patients because that concern is authoritative for them. That is what it means to be able to recognize and respond to a normative concern "on one's own," as a more-or-less capable, authoritative participant in the practice. Completing that transition, from novice imitator to participant competent to judge what responses would be appropriate in which situations in a practice, is itself a normative matter rather than a determinate matter of fact, even though it answers to evidence (Kukla 2000, 2002).

We can now see more clearly why both the predictive and the authoritative senses of normativity must work together. Initiation into a practice at first seems to place established participants in positions of unchallengeable authority over initiates, individually and collectively. They provide ostensive orientation to relevant differences in circumstances, expressive articulation of these differences and how they matter, and exemplify and guide response to those circumstances according to the practice's constitutive concerns. Any challenge to an instructor's or guide's authority can be overridden by the aligned responses of other established participants: "this is what we do," in the predictive sense. Initiates' failures to accord with those authoritative recognitions, performances, and assessments are then a failure to "get it" rather than a challenge to the practice and its constitutive concerns or to other participants' understanding of those concerns.

Here we encounter a more important reason why the predictive sense of normativity is never sufficient. It is not enough that the new initiate learn to recognize and accord with what participants in the practice normally do in various circumstances. The initiate must also be gripped by this recognition, so as to let that recognition also be authoritative over her own subsequent orientation, recognitions, and the responses they make possible. Initiates to a practice, both individually and in mutual alignment, then come to have countervailing authority that makes their training a mutually transformative relationship. Novices might only "recognize" the practice's normative concerns, in the stronger sense of letting them be authoritative for their own performance, if the predictive sense of those concerns were revised. Other participants may dismiss their revisionist responses as mistakes or as evidence of incompetence, negligence, or failure, but this assessment can cut both ways. An inability to bring enough new participants into accord with this assessment instead becomes a failure on the part of the practice to sustain its animating concerns.

The resulting ambiguity between mistakes, innovations, or reforms can arise in different ways. Established participants may try to understand how and why those new to the practice see or do things differently, in order to clarify and correct that revised response as misguided. Sometimes that strategy succeeds, and new participants come into closer accord with the predictive sense of its constitutive concerns. At least three other outcomes are possible, however. First, established participants might instead come to recognize for themselves, authoritatively, that what the initiate sees or does accords better with what the practice was *already* aspiring to.

Even for established participants, the predictive sense of normativity does not exhaust a practice's normative concerns, which always potentially outrun their own predictable determination or articulation. Second, however, since a practice can only be sustained if it gains new participants, previously established participants may have to take some initiates' alternative conceptions or performances into account predictively, acknowledging them as now normal or permissible variations in how the practice's constitutive concerns are recognized and realized. Third, they might instead take a stand against these revisionist normative concerns. That stance nevertheless risks the possible disappearance of the practice, so construed, if the normative concerns they take as constitutive cannot be sustained and taken up by others.

Recognizing that the authoritative sense of normativity has priority over its predictive sense explains not only why the latter is not fully reliable, but also why it cannot and should not be so. Some reasons are familiar for why people's acknowledgment or endorsement of a normative concern as authoritative is not fully predictive. People make mistakes, backslide, or override one concern in favor of others. The temporally extended relations among performances of a practice and its constitutive concerns nevertheless provide deeper reasons for a gap between the predictive and authoritative senses of normativity. Patterns of prior performance, even when they include discursive expressions of what the practice aims to accomplish, do not uniquely determine how that pattern should continue. People may confront and answer to the same normative concerns that already animate an ongoing practice, despite different conceptions of those concerns, how to respond to them, and how that response matters.

Such differences do not just arise from recruiting new participants. Participants may differ in their experience and skill, leading them to perform differently, with different rationales. Participants may also respond in novel ways to relevantly novel circumstances, and can disagree about whether circumstances are relevantly different. People can also be misidentified as participants responsive to a practice's normative concerns; who is or is not a participant in a practice is itself sometimes at issue. In these and other ways, the authority of normative concerns for participants in a practice can diverge from the predictive sense of normativity, even though the latter provides a baseline that constitutively enables its own authoritative reinterpretations. The interdependence of participants' performances nevertheless constrains those differences and brings the predictive and authoritative senses of normativity closer together in their implications. The distinctive, two-dimensional normativity of a practice-based way of life arises from the temporally extended interplay between these opposing tendencies.

The Temporality of Normativity

Practices are held together by the closely coupled interdependence of their performances. Participation does not involve doing what others do, but only requires that others' performances and relevant circumstances be sufficiently *aligned* to sustain their coupled interdependence. Misalignments nevertheless arise on a regular basis. People's participation in multiple practices may lead to conflicts. They have typically encountered different histories of past performance and different training and development, which affects their own performances. They may differ in how they understand or desire for a practice to proceed. Circumstances are also not always conducive to their intended performances. As a result, people often set themselves to recognize and respond to their situation in ways misaligned with what others do or with other circumstances that affect their performance. For these and other reasons, human ways of life are fraught with complications, contradictions, and conflicts. Encountering such misalignments, participants may adjust what they do, rearrange the circumstances, call for others to adjust, persevere in the face of incongruity or failure, or reconfigure their lives to abandon participation. Issues arise within a practice wherever performers encounter misalignments due to the resistance or crossed purposes of others or the recalcitrance of equipment or other circumstances. Responses to these issues can be misaligned in turn, whether due to differing recognition or conception of the issue, conflicting efforts at adjustment, or ineffective modifications of recalcitrant circumstances.

What does account for whether and how practices are sustained over time and how their performances are mutually accountable is participants' *need* for effective alignment with others in conducive circumstances. Most of what people do depends on such alignments, that is, on successful continuation of practices. Absent appropriate support, those performances would not be effective. Indeed the expectation of support—the predictive sense of normativity—is crucial to the intelligibility of those performances to the agents themselves or to others, even when that expectation is threatened by misalignments. Participating in an ongoing practice is not just doing what has been done before; it requires adjustments in response to changed circumstances, including changes in what others do. Participants thus answer to one another, and often try to hold one another accountable for such accommodations. Misalignments with others' performances or other circumstances thus motivate adjustments of what one does to reestablish supportive alignment, including efforts to induce others to adjust.

Practices consequently involve a constitutive dynamic tension. Out of a need to participate in some of the practices that make up their developmental environments, people adjust what they do in response to misalignments they encounter in order to sustain the practice and its constitutive goal-directedness. Others adjust what they do in turn. Each does so from only partial and partially overlapping acquaintance with the practice so far, amid differences in their own aspirations and needs. Thus, although everyone to some extent aims to adjust their performances and involvement to sustain sufficient alignment with others to sustain the practice's ends, that aspirational alignment is a moving target. Authoritative responses to misalignments may even directly counter some predictive aspects of past performance, taking the latter as telling indications of what *not* to do in light of the resulting misalignments. Even for such abrupt shifts in a practice, however, the force of that "not" arises from other continuities with past performances.

What initially seems at stake in resolving issues arising within a practice is whether the practice continues, including whether its characteristic concerns are recognized and satisfied, revised, or abandoned. Some responses to misalignments may lead others to modify or discontinue their participation, thereby affecting what is at issue for others. Practices only exist if current and subsequent generations continue to participate. If enough people stop acting in these ways, or act in sufficiently discordant ways, the practice falls apart or disappears. This fate has befallen languages that no one now speaks, occupations that no one takes up, sciences that no one investigates, and religious practices with no adherents. Not only does no one now undertake these practices, but they no longer *can* do so in the absence of other supporting performances or practices. Confronting that prospect, trade-offs thereby arise between *whether* a practice continues and what it is *becoming*, that is, its constitutive goal-directedness. Only some modifications would be acceptable to participants or effective in sustaining or transforming its concerns, and some require further adjustments elsewhere. Such trade-offs are a consequential evolutionary novelty in the hominin lineage. Other organisms develop and evolve, but their life-patterns are one-dimensionally normative. What is at stake is only *whether* their lives and lineage are maintained and reproduced. Human evolution has become two-dimensionally normative. What is at stake in how people respond to issues arising within their lives and life-patterns—the practices they take up and others on which they depend—is both *whether* their lives and lineage continue, and *how* they are lived. The latter concern encompasses which practices are sustained with which constitutive ends.

In discussing how participation in practices is accountable to what other participants do amidst other changing circumstances, I have spoken of normative "issues," "concerns," or "stakes," rather than norms, goods, virtues, rules, commitments, obligations, laws, or other determinate normative statuses. Normativity is often identified with accountability to determinate, authoritative norms. Understanding normativity as biologically grounded changes this order of explication. Biological normativity is not accountability to (already-)determinate norms, but instead concerns how various events and processes are dependent on and hence accountable to *one another* for continuation of their mutual interdependence. Each component of a life cycle "ought" to adjust to how others develop and evolve to sustain the process to which they each belong and upon which they each depend. The force of that "ought" comes from the prospect of death or extinction as the characteristic forms of biological failure. Similarly, the force of whether and how one ought to take up or continue in a practice comes from the prospect of its disappearance or its failure to sustain its constitutive ends.

Discursive practices do enable explicit articulation of normative issues or concerns and how they bear on a practice, but these expressions are themselves part of the practice. They do not simply *discern* a standard, virtue, or good that already governs or defines that practice. Such expressions help coordinate and guide what participants do, how they understand their own participation, and how the practice aligns with others. Efforts to articulate the issues and stakes in practices can nevertheless also be misaligned with one another or with what participants do. What people do often does not fit how they "make sense" of what they do. Other participants may also make different sense of the practices they take up together. Such misalignments may arise through mutual misunderstanding, misdirection, or disagreement, and their direction of fit cannot be settled in advance. Sometimes people adjust what they do to accord better with the norms, goods, or virtues proposed to express what the practice is up to. Sometimes these expressions are instead revised to accord better to what participants actually do.

Recognizing that talking about the normative interdependence of a practice's performances is part of the practice brings out the anaphoric character of such talk. The terms 'issues' and 'stakes' provide an explicitly anaphoric normative vocabulary. Participants in a practice—a temporally extended and spatially distributed pattern of interdependent performances—encounter the same issues, and answer to the same stakes. Participants' concern for whether and how that practice continues establishes the anaphoric accountability of its situated performances to one another for what is at issue and at stake in the practice. They confront the same issues and stakes despite different "conceptions" of what the issues are and what is at stake, because everyone's participation depends on how the practice as a whole develops over time.

Participants take up an already-ongoing pattern of practice both in response to its preceding patterns and oriented toward their continuation in perhaps modified forms. They have typically encountered only a partial, selective subset of prior performances and vary considerably in their grasp of the practice's development over time and in different settings. They do, however, normally understand others as engaged in activities that support, enable, and sometimes conflict with their own participation, and they rely on the anticipated subsequent continuation of sufficiently supportive roles by others in sufficiently conducive circumstances. Each understands and copes with their situation and its challenges and conflicts as encompassing not only one another's performances but also their *conceptions* of the situation they share and the *issues* it raises for them all. In responding to the situation by modifying what they do and inducing others to realign their performances, each participant redirects the future continuation of the practice towards accord with what each takes to be at issue and at stake. What the issues "really" are, however, and what is at stake in how those issues are resolved, always lies ahead of how they are variously conceived at any time.

Emphasizing the constitutively open-ended character of practices does not deny that the resulting patterns and institutions can be relatively stable or that expressions of their issues and stakes can be widely shared or endorsed. Practices are often sustained by achieving more-or-less settled resolution of issues they confront. Such resolutions are nevertheless also temporally extended projections that remain open to new issues. Issues may arise within an ongoing practice, from changed alignments with other practices, or from new participants for whom the practice is differently situated within their lives. Recognizing that what people do may belong together in normatively interdependent practices despite partial misalignments among their performances and assessments highlights a central problem for more traditional conceptions of practices as defined by performative regularities, shared presuppositions, or determinate governing norms. Achieving and sustaining a pattern of regular performance or of accepted presuppositions or norms cannot explain the continuity, intelligibility, and normative accountability of practices and their constitutive performances. Any such commonalities are instead consequential and sometimes fragile achievements in need of explanation.

Two-Dimensionality Revisited

The two-dimensional normativity of human ways of life originated with the evolution of a cooperatively interdependent way of life and its subsequent differentiation into multiple interdependent practices. One dimension of normative concern initially arises from within particular practices.

The closely coupled alignment of people's situated performances lets them undertake activities, develop skills, and achieve outcomes not possible without mutual support in conducive circumstances. The "goods" thereby achieved, that is, those practices' constitutive ends, are often only recognizable as such by those whose postures and skills have developed in contexts that make those goods salient and intelligible. Languages, economic exchange, food acquisition and preparation, religious worship, childrearing, manufacturing, scientific research, games, transportation systems, and the many other practices now overlapping in human ways of life each have their characteristic normative concerns. These include the skills and background understanding required to participate; the goods towards which they are directed, the hazards to be avoided, and the ability to recognize and appreciate both; the virtues and vices contributing to or detracting from effective participation; the suitability of materials, equipment, and settings; appropriate ways of timing, locating, coordinating, and rewarding or correcting one another's performances; and many other issues that motivate, sustain, or challenge the supportive alignments that sustain a practice. Sometimes these concerns are explicitly articulated or contested in terms whose significance and uses are part of the practice. Even then, however, those concerns and their discursive expression are also embedded in people's developed bodily postures and postured responses to one another's performances in various circumstances.

These proximate forms of practice-constitutive concern are only partially autonomous, however. One reason for their limited autonomy is the material circumstances they incorporate, which constrain how the practice evolves. Those circumstances may have been transformed by iterated cycles of niche construction, but those transformations were also dependent on their causal capacities. The emergence of *multiple* partially overlapping, interdependent practices is also crucial to their evolutionary significance in the human lineage, however. Although what participants do is proximately responsive to concerns internal to the practices they take up, their participation must also respond to their involvement in other practices, and how other practices bear on those proximate concerns and performances. These overlapping involvements are often what lead to adjustments in people's performances that partially misalign with what others do or with the surrounding circumstances. Misalignments can reverberate through multiple practices, because of how one practice depends in more limited ways on supportive alignment with others. Shifts in one practice that remove or change its supporting contribution to others may generate new issues in those practices as well, and the responses to those issues may then reverberate back or extend further outward.

The temporality of normativity as a biological phenomenon and its two-dimensional articulation in human ways of life have important consequences for how to think about critical assessment of practices and our ways of life. Over the course of evolution, countless organisms have lived and died, and myriad species and larger taxa have emerged and become extinct. Some emerged under propitious conditions and flourished before passing away. Others encountered straightened circumstances, more successful competitors, or were caught up in occasional mass extinctions. We can meaningfully assess them as having succeeded or failed to maintain life or lineage, including whether they developed and evolved to accommodate changing conditions, but there are no further plausible grounds for assessing their ways of life or their evolutionary adaptations as better or worse. In one respect, our lineage is no different. It emerged from common ancestry with other primates amidst stressful and rapidly shifting climatic changes in its east African habitats (Potts 1996). Within a geologically short span, that lineage dwindled to a single surviving species that nevertheless also migrated globally and massively expanded demographically. Sooner or later our species will pass on in turn, outlived or supplanted by others, or encountering or generating environmental challenges it cannot accommodate.

Extraordinary new possibilities for assessment and self-transformation nevertheless emerged *within* our ways of life. The niche constructive diversification of human ways of life into multiple interdependent practices enables trade-offs between which practices are sustained and how those practices change. The closely coupled interdependence of a practice-differentiated way of life is more readily open either to rapid and significant transformation or to stabilizing responses. New degrees of freedom and new possibilities for assessing and changing one's situation thereby emerge within the ongoing reproduction of those ways of life. The emergence of language and other conceptually articulated practices—these include ostension, images, music, dramatic impersonation, equipmental complexes, and more—dramatically expanded those degrees of freedom by enabling participants in practices to express issues the practice confronts, and hold others to account for performances that fail to align with others.[2]

Perhaps the most striking transition, however, is that the constitutive goal-directedness of our way of life is no longer limited to survival of the lineage. Practices depend on one another in ways that open a further range of normative concerns for *how* various practices fit together within individual lives and our encompassing, practice-differentiated way of life. Familiar examples of such integrative concerns include social justice, participatory governance, collective power or achievement, individual freedom, epistemic insight, personal ties of kinship and affiliation, or religious commitments. In this more complex field of goal-directedness and critical assessment, people do not just act either to advance normative concerns internal to particular practices or else to maintain life and lineage. They are also responsive to issues raised by how the practices they take up depend on, support, and answer to one another. What is at stake in resolving the new kinds of issue that human individuals and populations thereby confront is not only whether our lives and ways of life continue, but what we are thereby becoming. Many of us conclude that what is most crucially at stake in *our* practice-differentiated way of life is whether it fulfills one or more of these integrative normative concerns, such as for justice or limits on our environmental footprint. It is also no longer so obvious, however, whether the emergence of a practice-differentiated, discursively articulated way of life really was a biologically successful adaptation, or was instead a spectacular but relatively short-lived aberration. That discussion, however, must be reserved for another occasion, and perhaps for pending empirical tests.

[2] For discussion of the sense in which these other practices are "conceptually" articulated, see Rouse (2015, 157–60; 2023, Ch. 7).

References

Bourdieu, Pierre. 1990. *The Logic of Practice*. Translated by Richard Nice. Stanford: Stanford University Press.

Brandom, Robert. 1979. Freedom and Constraint by Norms. *American Philosophical Quarterly* 16: 187–196.

Giddens, Anthony. 1984. *The Constitution of Society*. Berkeley: University of California Press.

Gilbert, Scott. 2016. Developmental Plasticity and Developmental Symbiosis: The Return of Eco-Devo. *Current Topics in Developmental Biology* 116: 415–433.

Kukla, Quill Rebecca. 2000. Myth, Memory and Misrecognition in Sellars' 'Empiricism and the Philosophy of Mind'. *Philosophical Studies* 101: 161–211.

———. 2002. The Ontology and Temporality of Conscience. *Continental Philosophy Review* 35: 1–34.

Odling-Smee, John, Kevin Laland, and Marcus Feldman. 2003. *Niche Construction*. Princeton, NJ: Princeton University Press.

Potts, Richard. 1996. *Humanity's Descent*. New York: Avon.

Rouse, Joseph. 2015. *Articulating the World*. Chicago: University of Chicago Press.

———. 2023. *Social Practices as Biological Niche Construction*. Chicago: University of Chicago Press.

Sterelny, Kim. 2012. *The Evolved Apprentice*. Cambridge, MA: MIT Press.

Sultan, Sonia. 2015. *Organism and Environment*. Oxford: Oxford University Press.

Taylor, Charles. 1995. To Follow a Rule. In *Philosophical Arguments*, 165–180. Cambridge, MA: Harvard University Press.

Turner, Stephen. 2007. Practices Then and Now. *Human Affairs* 17: 111–125.

Wittgenstein, Ludwig. 1953. *Philosophical Investigations*. Translated by Elizabeth Anscombe. Oxford: Blackwell.

Avoiding "Tack-on" Theories of Culture

Kevin M. Cahill

Introduction

It is a commonplace that some broad sense of "the social" is central to Wittgenstein's later writings, including those writings devoted to mathematics. By "the social" I mean a nest of ideas connected to the learning, teaching, and performance environment of symbolic activity. Although there might be disagreement as to precisely how to assess their significance, most commentators would agree that notions such as practice (*Praxis*), use (*Gebrauch*), custom (*Gepflogenheit*), technique (*Technik*), and form of life (*Lebensform*) are important for understanding Wittgenstein's later thought. But this widespread agreement about the significance of the social for Wittgenstein's later philosophy could obscure an important distinction within it. It could obscure a distinction between what I will here refer to as "anthropological naturalism" from "sociological naturalism". Support for both of these viewpoints can be found in

K. M. Cahill (✉)
Department of Philosophy, University of Bergen, Bergen, Norway
e-mail: kevin.cahill@uib.no

K. M. Cahill (ed.), *Wittgenstein on Practice*,
https://doi.org/10.1007/978-3-031-68656-6_13

Wittgenstein's writings, sometimes in the same passage. The distinction I have in mind can be roughly described as follows:

1. Sociological naturalism often implies a strong empirical element, encouraging finding answers to philosophical questions through investigation of the various practical and theoretical approaches taken by workers in various pursuits. The main question then becomes "What do mathematicians, scientists, doctors, engineers, police, carpenters, or tailors generally say or do when they carry out their respective activities?" Some things Wittgenstein says (or is recorded as having said) can be read as pointing in this direction.[1]
2. Anthropological naturalism addresses the question of what kind of creatures are involved in such symbolic endeavors. Here, looking at practices, customs, and techniques may be regarded as important for *revealing* aspects of, even if not necessarily explaining, the intelligibility of symbolic life. Passages depicting elementary learning situations, such as Augustine's recollection of learning the meanings of words as a child (*PI* §1), ostensive definition (*PI* §29 & §257), or of course the "wayward pupil" in the remarks on rule-following (*PI* § 143) are from this perspective best read as invitations to imagine how we as agents are brought to relate to such symbolic contexts.[2]

It is reasonable to see both kinds of naturalism figuring in Wittgenstein's writings. Yet an exclusive focus on the former can mislead philosophers into devoting their attention solely, for instance, to the practices of working professional mathematicians or scientists.[3] After all, if Wittgenstein enjoins us to "look and see" how a term is used if we want to grasp its

[1] See, for example, the references to tailors in Ludwig Wittgenstein, *Lectures & Conversations on Aesthetics, Psychology, and Religious Belief*, ed. Cyril Barrett (Berkeley: University of California Press, 1966) 13. This description may also be read as an example of "anthropological naturalism." Many of the relevant remarks share this duality.

[2] Ludwig Wittgenstein, *Philosophical Investigations*, eds. G.E.M Anscombe & Rush Rhees (Oxford: Blackwell, 1953 (1997). Henceforth 'PI'.

[3] For an account of scientific knowledge that purports to draw on Wittgensteinian insights into practice, see David Bloor and Barry Barnes, *Scientific Knowledge: A Sociological Approach* (Chicago: University of Chicago Press, 1996).

meaning (*PI*, §66), what better way to understand the meaning of, say, "number" than to look and see how mathematicians operate with this term? This approach is in line with a widespread and now many-decades old naturalist trend in philosophy more generally, wherein one is supposed to eschew mere "armchair" speculation and take one's problems and solutions mainly, if not exclusively, from the activities of experts in the respective fields.

I think that this tendency in the philosophy of the sciences, including the philosophy of mathematics, has been largely for the good. In the case of assessing Wittgenstein's thought, however, too much focus on higher-level cognitive practices such as sophisticated theories, proofs, or even experimental activity risks bypassing something else he is frequently targeting. For it can be very far from obvious how Wittgenstein's frequent discussions of extremely elementary linguistic and mathematical situations are supposed to shed light on the technical questions that are sometimes taken to be the very stuff of mathematics or logic. I think Wittgenstein himself was acutely aware of this apparent difficulty. Thus, on the very first pages of the student lecture notes taken during his 1939 lectures on the foundations of mathematics we find passages such as the following:

> I am proposing to talk about the foundations of mathematics. An important problem arises from the subject itself: How can I—or anyone who is not a mathematician—talk about this? What right has a philosopher to talk about mathematics? (Wittgenstein 1976, 13)

> Again, one might think that I am going to give you, not new calculations but a new interpretation of these calculations. But I am not going to do that either. I *am* going to talk about the interpretation of mathematical symbols, but I will not give a new interpretation. (Wittgenstein 1976, 13)

> I can as a philosopher talk about mathematics because I will only deal with puzzles which arise from the words of our ordinary language, such as "proof", "number", "series", "order", etc. (Wittgenstein 1976, 14)

> [A]ll the puzzles I will discuss can be exemplified by the most elementary mathematics—in calculations which we learn from ages six to fifteen, or in what we easily might have learned, for example, Cantor's proof. (Wittgenstein 1976, 14)

Especially with regard to the philosophy of mathematics, a failure to appreciate the importance of Wittgenstein's extended discussions of simple symbolic situations, combined with the claim that Wittgenstein did not understand much higher mathematics can be used to downplay his relevance for this branch of philosophy.[4] No doubt Wittgenstein's reasons for paying so much attention to simple examples are of a piece with his preference for applications in mathematics in particular, and language in general. Be that as it may, failure to distinguish anthropological naturalism from sociological naturalism makes it hard to assess Wittgenstein's work fairly, especially on mathematics, because it fails to recognize that his targets were typically conceptually prior to, or at least different from, more technical issues having to do with higher mathematics and logic. Gaining more understanding of these issues will, I hope, clarify how we might read Wittgenstein's later work as allowing us to avoid "tack-on" theories of culture.

In section "Why the Focus on Everyday Language?", I look at the significance of Wittgenstein's employment of simple scenarios from ordinary language, suggesting that a concern for the issue of radical skepticism accounts in large measure for the prominence of these examples in his writing. In section "Practices, Customs, Uses", I discuss the ideas of use, custom, and practice in Wittgenstein's later thought. Relying on the work of John McDowell, I bring out how we should not construe these concepts in Wittgenstein. In section "Culture and Evolution", building on ideas from McDowell, I engage in some hopefully non-metaphysical speculation about the relation between cultural and biological evolution, and about how this relation touches on the question of how to avoid tack-on theories of culture. I develop these ideas with the aid of some neglected writings of Clifford Geertz.

---

[4] On this see Juliet Floyd, "A Note on Wittgenstein's 'Notorious Paragraph' about the Gödel Theorem" (with Hilary Putnam), *Journal of Philosophy* 45, 11 (2000): 624–632.

Why the Focus on Everyday Language?

One crucial respect in which Wittgenstein's anthropological naturalism and exclusively sociological approaches pass one another by is that whereas an interest in what is at stake in radical or "meaning" skepticism is often central for the former, it is usually not a concern for the latter.[5] By "meaning skepticism" I mean a problematic in which it appears that our very handle on the correlative concepts of meaning and understanding seems to evaporate, in contrast with epistemological concerns about the possibility of knowledge. This problem is of course an essential feature in the dialectic of the remarks on rule-following, something that Saul Kripke admirably revealed.[6] In fact, radical skepticism has been an actual *consequence* of certain exclusively sociological readings of Wittgenstein, largely because these easily degenerate into versions of behaviorism that blot out any genuine possibility of correctness altogether.[7] Against this, an anthropological naturalist reading of Wittgenstein will see the running motif of radical skepticism in his work as precisely the sort of puzzle that his employment of concepts such as custom, practice, and form of life are meant to help us transcend, not solve.[8] Sociological naturalist readings that enjoin us to just "look at the practices" of scientists or mathematicians, are, while potentially instructive as

---

[5] By "radical skepticism" I am pointing to a distinction between what James Conant has termed "Kantian" as opposed to "Cartesian" skepticism. See James Conant, "Two Varieties of Skepticism," in *Rethinking Epistemology*, ed. Guenter Abel and James Conant, vol. 2 (Berlin: Walter De Gruyter, 2012), 1–73. In short, "Cartesian" skepticism involves the question "How can I ever know my thoughts or ideas correspond to a truth independent of my own consciousness?". "Kantian" skepticism, on the other hand, is directed at the even more troubling (pseudo)-question, "How can I ever know if my thoughts make any sense?" There may be problems with this schematic presentation, but I can't enter into them here.

[6] See Saul Kripke, *Wittgenstein on Rules and Private Language: An Elementary Exposition* (Oxford: Blackwell, 1982).

[7] Here I have in mind especially Bloor and Barnes. In a sort of ironic twist, however, after revealing the paradox of meaning in the remarks on rule-following, and thus generating an enormous literature on radical meaning skepticism, Kripke himself attributes an unsatisfying version of sociological naturalism to Wittgenstein, a "solution" that merely exacerbates the problem.

[8] This point is of course central to the work of Stanley Cavell and John McDowell. Cavell's repeated claim that the temptation to radical skepticism is bound to recur in the course of human life is, however, something that requires a view of human nature that I have criticized in Ch. 3 of my *Towards a Philosophical Anthropology of Culture: Naturalism, Relativism, and Skepticism* (New York: Routledge, 2021).

far as they go, apt to suggest that Wittgenstein is offering constructive solutions to philosophical problems, despite his warnings to the contrary (*PI* §109 & §124). For his purposes, these approaches overlook the facts of meaning and understanding presupposed by the sciences, and so risk coming in too late.

Consider first an evocative remark from early in *Philosophical Investigations*:

> [A]sk yourself whether our language is complete;—whether it was so before the symbolism of chemistry and the notation of the infinitesimal calculus were incorporated in it; for these are, so to speak, suburbs of our language. (And how many houses or streets does it take before a town begins to be a town?) Our language can be seen as an ancient city: a maze of little streets and squares, of old and new houses, and of houses with additions from various periods; and this surrounded by a multitude of new boroughs with straight regular streets and uniform houses. (*PI* §18)

I take this passage as my starting point in order to develop what I consider to be one of Wittgenstein's central reasons for focusing on the practice of ordinary language in so much of his later work. The remark suggests to me that from one perspective at least, the "ancient city" of "our language" can be thought of as having a kind of completeness to it before the addition of "suburbs" such as chemistry or calculus. But what reasons are we given for regarding the ancient city as primary in any but a temporal sense? This is not merely a genetic point, but also a logical one, with implications for Wittgenstein's handling of radical skepticism.

One reason why simple symbolic activities are so central to Wittgenstein's later thought concerns the way in such activities share a kind of internal connection to the all-important idea of normativity, the possibility of performing a symbolic activity, or a symbolically infused activity, correctly.[9] Whether the issue concerns the meaning of an expression or, correlatively, the attribution to someone of understanding the

[9] By a "symbolically infused activity" I am referring to something like carpentry, which, although not strictly speaking symbolic, takes place for creatures like us who share a discursively articulated environment.

meaning of an expression, without some default notion of seamless, immediate orientation towards language and the world, the very concept of correctness, and with it meaning, disintegrates altogether. By contrast, philosophers and other theorists who uncritically adopt the attitude that language consists firstly of inert chirps or marks in need of an intellectually sophisticated and detached act of interpretation (i.e., theory) in order to become meaningful, lose their handle on the very phenomenon of meaning altogether.[10] This theme is pervasive in Wittgenstein's later writings.[11] He characterizes this assumption at *PI* §431 this way:

> "There is a gulf between an order and its execution. It has to be filled by the act of understanding."
>
> "Only in the act of understanding is it meant that we are to do THIS. The *order*—why, that is nothing but sounds, ink-marks."

And, of course, earlier at *PI* §201, one of the culminating points of the skeptical dialectic on rule-following, readers will already have encountered Wittgenstein's rejection of the assumption that our logically prior relation to language is to uninterpreted strings such as '+2'.

> This was our paradox: no course of action could be determined by a rule, because every course of action can be made out to accord with the rule. The answer was: if everything can be made out to accord with the rule, then it can also be made out to conflict with it. And so there would be neither accord nor conflict here.
>
> It can be seen that there is a misunderstanding here from the mere fact that in the course of our argument we give one interpretation after another; as if each one contented us at least for a moment, until we thought of yet another standing behind it. What this shews is that there is a way of grasping a rule which is *not an interpretation*, but which is exhibited in what we call "obeying the rule" and "going against it" in actual cases.

---

[10] For an argument that this anti-skeptical attitude pervades Wittgenstein's early work, see my "The *Tractatus* and the Carnapian Conception of Syntax" in *The Tractatus at 100*, eds. Martin Stokhof and Hao Tang (London: Palgrave Macmillan, 2023).

[11] For recent work on this issue see James Conant, "Wittgenstein's Critique of the Additive Conception of Language," *Nordic Wittgenstein Review*, vol. 9 (2020): 7–36.

The primacy of an immediate, non-evidential relation to the expressions of ordinary language is also a major theme that Wittgenstein voices in *On Certainty*. In one of many references to G.E. Moore's argument for the existence of the external world, which relied on Moore's taking the truth of the propositions asserting the existence of his own hands as being based on indubitable visual evidence, Wittgenstein points out that before there can be evidence of any kind for the empirical truth of a proposition, there is, on pain of the "abyss" of meaning skepticism, an unmediated understanding of the sense of the words in which certain judgments are given to us in the first place. Thus, we find at *OC* §370,

> The fact that I use the word "hand" and all the other words in my sentence without a second thought, indeed that I should stand before the abyss if I wanted so much as to try doubting their meanings—shews that absence of doubt belongs to the essence of the language-game, that the question "How do I know…" drags out the language-game, or else does away with it.[12]

One upshot of the above considerations is that ordinary language and the know-how which infuses it is fundamental (even if not foundational) to the epistemological perspective characteristic of theorizing. In other words, the conceptual dependency relation of ordinary language practices and those of theoretical activities such as the natural sciences and mathematics is not symmetrical. The simple fact is that ordinary language permeates the intelligibility of every form of scientific activity much more pervasively than the other way around. While we can pretty well conceive of people carrying on their lives with no science and very little theory, it's doubtful we could conduct any science at all without the orientation provided by ordinary language. Ordinary language can get along much better and longer without any theoretical stance, scientific or otherwise, than theorizing can get along without ordinary language. That does not demonstrate the dependence of an a priori necessity (assuming we knew what that was supposed to mean anyway), but it does strongly

[12] See also *OC* §126.

suggest a conceptual dependence whose reversal we would, to put it mildly, have a very hard time fathoming.[13]

So far I have focused on a distinction between the relation that competent language users typically have to ordinary language versus the more detached, reflective attitude taken in learning to respond to inference based assertions or theoretical constructions.[14] But simply pointing out this atemporal difference between ordinary languages and theoretical languages omits something else essential to Wittgenstein's discussions, viz., the context in which children are brought to develop their sensitivities to symbolic life. At *On Certainty* §374, Wittgenstein elaborates and augments the same criticism of the theoretical view we just encountered above at §370, but now he connects it to the very teachability and learnability of language:

> We teach a child "that is your hand", not "that is perhaps (or "probably") your hand". That is how a child learns the innumerable language-games that are concerned with his hand. An investigation or question, 'whether this is really a hand' never occurs to him. Nor, on the other hand, does he learn that he *knows* that this is a hand.

We have seen evidence above that Wittgenstein's interest in ordinary language is essentially linked to his interest in revealing the incoherence of radical meaning skepticism. This same point is amplified in his interest in language learning.

Kripke's discussion of rule-following in the *Investigations* probably provides the best known example of the philosophical importance that Wittgenstein attaches to symbolic learning. This series of remarks nominally begins at §143.[15] But in fact the question of learning is already indirectly present much earlier.

---

[13] See *OC* §114–115 and §204–205.

[14] I am not arguing for a substantive, once-and-for-all line between ordinary language and the languages of theory. The distinction is contextual, functional, and phenomenological. For instance, many of us can arrive at a kind of fluency with certain relatively simple systems such as elementary arithmetic or relatively simple theories such as Euclidean geometry that can seem to rival our native tongue in immediacy.

[15] See PI §143–144.

> When they (my elders) named some object, and accordingly moved towards something, I saw this and I grasped that the thing was called by the sound they uttered when they meant to point it out. Their intention was shewn by their bodily movements, as it were the natural language of all peoples: the expression of the face, the play of the eyes, the movement of other parts of the body, and the tone of voice which expresses our state of mind in seeking, having, rejecting, or avoiding something. Thus, as I heard words repeatedly used in their proper places in various sentences, I gradually learnt to understand what objects they signified; and after I had trained my mouth to form these signs, I used them to express my own desires. (*PI* §1)

Augustine's recounting of how he learned the names of objects is apparently innocent enough, but it also reveals a source of potential philosophical confusion.[16] As Wittgenstein employs it, Augustine's recollection urges on us a picture wherein genuine name–object relations appear to be on display in the "brute" linguistic behavior of the adults whom the child Augustine observes. Without the given of grammatical "stage-setting," however, it is entirely unclear how the "Augustinian picture" of using a word to name an object is supposed to work.[17]

> Naming is so far not a move in the language-game—any more than putting a piece in its place on the board is a move in chess. We may say: *nothing* has so far been done, when a thing has been named. It has not even *got* a name except in the language-game. (*PI* §49)[18]

Unless we can assume some prior contextual understanding of the adults' actions, the utterances they make when they approach various objects might be taken to mean anything or nothing.[19] The problem of meaning skepticism is thus inchoate in the very first remark of the book.

[16] Whether what Wittgenstein says in his criticism of the "Augustinian picture" is fair to the broader meaning of what the man Augustine of Hippo writes in his recollection is not germane to my discussion here.

[17] Cf. PI §257. See also PI §29.

[18] Although it could seem as though Wittgenstein has moved on to new topics by PI §49, I am taking this remark to be still fundamentally in dialogue with issues already raised at PI §1.

[19] Naturally, unless we assume some correlative understanding on the part of the adults, it would be difficult to characterize their behavior as actions. But that is another matter.

Without recognition of our immersion in a web of significance, something that is part and parcel of our development as speakers and judgers within a community, it is obscure how we are supposed to grasp an act of naming an object, let alone correctly evaluate a full-blown assertion.[20] Wittgenstein remarks in this vein,

> If language is to be a means of communication there must be agreement not only in definitions but also (queer as this may sound) in judgments. This is agreement not in opinions, but rather in form of life. (*PI* §241–242)[21]

There are three things to note at this point. First, there is no indication by Wittgenstein here or anywhere else that such agreement, such attunement (*Übereinstimmung*), should be wholly ascribed to our common biological inheritance (though there is no reason, either, to doubt its necessity). On the contrary, we are fully *brought into* this attunement in judgments when we learn the practices of rendering them. The temporal tenor of the two following couplets from *On Certainty* makes this cultural-historical strain in Wittgenstein's thought on this matter pretty hard to avoid:

§128. From a child up I learnt to judge like this. *This is* judging.
§129. This is how I learned to judge; *this* I got to know *as* judgment.
§139. Not only rules, but also examples are needed for establishing a practice. Our rules leave loop-holes open, and the practice has to speak for itself.
§140 We do not learn the practice of making empirical judgments by learning rules: we are taught *judgments* and their connexion with other judgments. *A totality* of judgments is *made* plausible to us. [my italics with 'made'.][22]

---

[20] I do not mean to invoke Quine's notion of a "web of belief" here. Wittgenstein is not a holist in that sense. See Ch. 1 in *Towards a Philosophical Anthropology of Culture.*

[21] I have reversed the order of these remarks. See also *OC* §128 and §140.

[22] Lest anyone think that the significance of practice is restricted to the case of making empirical judgments, we should recall that Wittgenstein also tells us at PI §201 that "obeying a rule is a practice."

Second, there is a clear holistic thrust here: "form" as in "form of life" resists by definition any atomistic interpretation. Wittgenstein's more contextual, holistic perspective on language differs fundamentally from compositional semantics. Of course when we are considering a language one already understands, the latter can be regarded as a plausible commonsense view of sentence meaning: *of course* a prior familiarity with the logical features of words from earlier contexts might enable one to see the meaning of a sentence as in some manner "composed" out of the individual constituent meanings of those words. But when this idea is forwarded as a *de novo philosophical* explanation of our understanding of sentence meaning, it simply doesn't work. It clearly presupposes that one has somehow *already* managed to garner that familiarity from those prior contexts, and this familiarity is simply *not* available within the objectified and atomistic confines of the Augustinian picture of learning.[23] This is Wittgenstein's point at *PI* §32 when he says,

> Augustine describes the learning of human language as if the child came into a strange country and did not understand the language of the country; that is, as if it already had a language, only not this one.

Third, it is worth noting that there is no reason for thinking that Wittgenstein believes that the salient features of our immediate relations to symbolic life are absent in more theoretically complex, cognitively advanced contexts. Clearly, for example, scientists discuss highly theoretical matters amongst themselves within a framework provided by ordinary language. Wittgenstein stresses simple learning cases, therefore, because these are most apt for displaying normatively relevant characteristics that run through our entire symbolic life from childhood to adulthood, in particular our immersive orientation to ordinary language. One might say that in learning a theory, we learn symbolic responses of a "higher" sort, whereas in the simple learning cases Wittgenstein often describes, what is at stake is our symbolic responsiveness *überhaupt*. And it goes without saying that we learn our first language(s) from others, immersed

---

[23] See again *PI* §241–242. See *OC* §128 and §140. This question of holism vs atomism goes back to the very origins of analytic philosophy. See Peter Hylton, *Russell, Idealism, and the Emergence of Analytic Philosophy* (Oxford: Clarendon Press, 1990).

in a cultural-biological form of life. This capacity to learn a natural language is no doubt underwritten and shaped by various general facts of our common human nature. Still, our symbolic sensitivities come to be fully developed in our acquiring language(s), which just so happen to be isomorphic with the particular natural language(s) we learn from others in a sociocultural setting. We humans don't learn to speak "human"; we learn Urdu, Spanish, Dutch, etc. The human form of life is one of plurality. I will return to this point briefly in section "Culture and Evolution".

I take some conclusions here to be that everyday language for Wittgenstein is *not* a theory (and therefore not "folk psychology") and that humans do not learn their first language, what later becomes the bulk of the body of ordinary language, either *as* a theory *or* in virtue of another, tacitly held, theory. Or, to state the matter less dogmatically, to imagine that one can seriously entertain that these might be the case is to introduce radical unintelligibility into the very heart of language and thought (and so into the same theories that lead to this nonsense in the first place). Putting it bluntly, we simply have "no model" of what it would be like to construe our understanding of our first language in terms of the usual epistemological categories of an individual subject representing objects in terms of a theory and arriving at conclusions through the use of a sophisticated system of evidence gathering and inferential relations. Wittgenstein's philosophical distance to the idea that the immediacy of the everyday "lifeworld" might be intelligibly eliminated by theory cannot be overstated. But he is not offering an explanation of normativity in terms of the practices of ordinary language, but highlighting a recalcitrant fact of great conceptual import. But as we'll see next, there are better and worse ways to imagine the strong connection between correctness and the social.

Practices, Customs, Uses

Philosophical Investigations §198–§202 mark a key stage of the dialectic on rule-following. And in the course of these remarks Wittgenstein invokes the essentially social notions of training (*Abrichtung*), regular use (*einen ständigen Gebrauch*) of signposts, custom (*Gepflogenheit*), institutions (*Institutionen*), mastery of a technique (*eine Technik beherschen*), and of course practice (*Praxis*). Like so much in Wittgenstein's later philosophy, these concepts pertain to the question of the correct application of a sign, that is, symbolic normativity.

Many sympathetic commentators see Wittgenstein as having employed these and related concepts to mount a devastating attack on the idea that our understanding of rules must be accounted for by some fantastical mental act of interpretation. According to John McDowell, Wittgenstein does so by drawing out a confusion rooted in our insistence that a correct interpretation of the sign for an arithmetical rule be conceptually independent, both of those responses that we have previously accepted as being in accord with the rule, and of all the further possible moves we could give. When we are beholden to the requirement on rationality forced on us by such a mythology, acknowledging the conceptual interdependence of our understanding of the rule and our actual responses to it will at best seem like a compromise that practical necessity extracts from us; it will not appear to us to be in accord with rationality per se. Yet as McDowell noted in a now classic 1981 paper,

> [T]he idea that the rules of a practice mark out rails traceable independently of the reactions of the participants is suspect even in this apparently ideal case (a numerical series): and insistence that wherever there is going on in the same way there must be rules that can be conceived as marking out such independently traceable rails involves a misconception of the sort of case in which correctness within a practice can be given the kind of demonstration we count as proof.[24]

Nevertheless, mere recognition of the primacy of our socially inflected natural reactions does not prevent serious misunderstanding of Wittgenstein's aims.

Some behaviorist-oriented readers have found the presence in his work of ideas such as use, custom, and practice as providing the normatively inert materials for constructing a philosophical explanation of

---

[24] McDowell, 'Non-Cognitivism and Rule-Following', 146. McDowell's formulation here echoes PI §218. Elsewhere he refers to the mythology as the "view from sideways on" (McDowell 1994, 34–36) or the "Myth of the Disembodied Intellect" (McDowell 2007a, 349).

normativity itself.[25] More sophisticated interpreters have taken the later Wittgenstein's work to underwrite a view whereby meaning and normativity are somehow *undergirded* or created by practices into which we become acculturated through our upbringing and social training. Here, practices and related notions tend to be conceived as occupying a middle ground between mere instinctive, quasi-mechanical reactions à la behaviorism, on the one hand, and full-blown rationality often associated with higher cognitive functions, on the other.[26] On some of these accounts, practices in Wittgenstein's later work, then, are seen to function as quasi-transcendental conditions of the possibility of normativity, including everyday symbolic activities.[27]

We find a well-known example of this interpretation in the work of Hubert Dreyfus. In addition to his groundbreaking exegesis of Heidegger's *Being and Time*,[28] Dreyfus was also hugely influential in philosophical and popular discussions of both strong artificial intelligence and cognitive science.[29] This other work drew not only on the accounts of practice and action that Dreyfus found in existential phenomenologists Heidegger and Merleau-Ponty, but also on Dreyfus' reading of the later Wittgenstein. In fact, in his efforts to explain these Continental philosophers to analytic

[25] Kripke and Bloor and Barnes are important examples of such readings.

[26] In a 1984 paper, McDowell himself suggested that there was some kind of under-described Wittgensteinian program of practice theory yet to be fleshed out of his writings.

[27] This type of approach is not confined to philosophy. Pierre Bourdieu has argued that agency is best understood as dynamic, context-sensitive, and fundamentally grounded in a set of unconscious dispositions and sensibilities directly inscribed in the body of the agents comprising a given society. His well-known term for this set of dispositions is "habitus." Although Bourdieu once described Wittgenstein as "probably the philosopher who has helped me most at moments of difficulty. He's a kind of saviour for times of great intellectual distress," it's unclear to what degree his *Outline of a Theory of Practice* was actually influenced by Wittgenstein. At any rate, Bourdieu's habitus certainly looks like an attempt to employ ideas from the later Wittgenstein to provide a practice-based account for all intelligibility. For an explicitly Wittgenstein-inspired discussion in this vein, see Theodore Schatzki "Practices and Actions: A Wittgensteinian Critique of Bourdieu and Giddens," *Philosophy of the Social Sciences* 27: 283–308. For Bourdieu's description of Wittgenstein's influence, see "Fieldwork in Philosophy," Interview with A. Honneth, H. Kocyba, and B. Scwibs, Paris 1985, in (Pierre Bourdieu 1990. *In Other Words: Essays Towards a Reflexive Sociology*, 9. Stanford: Stanford University Press).

[28] See Hubert L. Dreyfus, *Being-In-The-World: A Commentary on Heidegger's Being and Time, Division I* (Cambridge: MIT Press, 1991).

[29] See Hubert L. Dreyfus and Stuart Dreyfus, *Mind Over Machine* (New York: Free Press, 1986), and Hubert Dreyfus, *What Computers Still Can't Do* (Cambridge: MIT Press, 1992).

audiences, Dreyfus frequently made use of arguments drawn from Wittgenstein. Thus, in his 2005 APA Pacific Division Presidential Address, Dreyfus wrote,

> [a]s Heidegger and Merleau-Ponty (as well as Wittgenstein) have argued, we can only relate to objects and make judgements about them insofar as they show up on the background of the world – and the world is not a belief system but is opened to us only through our unthinking and *unthinkable* engaged perception and coping.[30]

This was the opening salvo of a debate between Dreyfus and McDowell on the role of conceptuality in action, in which the nature of practice played a central role.[31] The debate mainly concerned issues about how we respond appropriately to solicitations from our social environment.[32] (It might, however, just as well have been about our receptivity to the promptings of ordinary language.) Dreyfus assumes that whatever is open to conceptuality, whatever is thinkable or sayable, must rest on a more fundamental ground floor of something nonconceptual, in this case the coping practices through which humans acquire the ability to navigate through their world. No doubt finding inspiration in Wittgenstein's talk of following a rule "blindly," he believes this ground floor is necessary to avoid a regress of justifications that is opened up as soon as this ground floor is contaminated with the least "bit" of conceptuality or thought.[33]

In Dreyfus' picture, coping practices function as the needed substratum for the normativity evidenced in paradigmatic meaningful activities, presumably including elementary rule-following. These practices,

---

[30] Hubert Dreyfus, "Overcoming the Myth of the Mental: How Philosophers can Profit from the Phenomenology of Everyday Expertise" (APA Pacific Division Presidential Address 2005) *Proceedings and Addresses of the American Philosophical Association* 79, no. 2 (November 2005): 58–59.

[31] This exchange originally appeared in *Inquiry*, Vol. 50, No. 4, August, 2007 as "What Myth?" (McDowell), "Return of the Myth of the Mental" (Dreyfus), "Response to Dreyfus" (McDowell), and "Response to McDowell" (Dreyfus).

[32] For the purposes of my discussion I will not address the differences between affordances and solicitations.

[33] PI §219. For an excellent paper which was written slightly before this debate but presages many of these issues, see Denis McManus, "Rules, Regression and the 'Background': Dreyfus, Heidegger and McDowell," *European Journal of Philosophy*, Vol. 16, No. 3, December, 2008, 432–458.

whether they are conceived of as inherently normative or somehow normatively neutral is not always clear to me, serve as a kind of quasi a priori ground for the linguistic and conceptual levels that are imagined as resting on them, as, so to speak, logically distinct lower and higher layers in a cake. The upshot here is that on pain of a regress of justification, normativity, correctness of response to what a situation demands, requires that this lower layer not be conceptualizable at all. A substantive binary question is thus apparently forced on us: "Do ordinary language and conceptuality necessarily rest on a base-line of embodied coping practices?" Dreyfus seems to think that Wittgenstein should be read as confidently answering "Yes, they do."

Whatever else one may think about it, the claim that rationality and normativity require grounding in non-conceptualizable coping practices certainly sounds very much like a piece of constructive philosophy. At least as far as interpreting Wittgenstein is concerned, passages such as *PI* §109 make evident, to me at any rate, that something has gone seriously wrong with Dreyfus's account:

> It was true to say that our considerations could not be scientific ones. It was not of any possible interest to us to find out empirically 'that, contrary to our preconceived ideas, it is possible to think such-and-such'—whatever that may mean….And we may not advance any kind of theory. There must not be anything hypothetical in our considerations. We must do away with all *explanation*, and description alone must take its place.[34]

Specifically in response to Dreyfus and in line with his "quietist" reading of Wittgenstein, McDowell counters that,

> Phenomenological attention to embodied coping should not be conceived as Dreyfus conceives it—as a way to answer the question "how the nonconceptual given is converted into a given with conceptual content." That question should be rejected, not answered.[35]

[34] See also, for example, *PI* §128 and §133.

[35] McDowell, 'What Myth?', 349. See John McDowell, 'Wittgensteinian Quietism', *Common Knowledge* 15:3 (Durham: Duke University Press, 2009) 365–372. McDowell's assertion that, "That question should be rejected, not answered" echos §189 in *Philosophical Investigations*.

Exegesis aside, as I see things, there are at least two philosophical problems with the ideas expressed by Dreyfus, although they may amount to the same problem. The first problem with taking practices, or anything else, to be the general conditions of the possibility of language, thought, or intelligibility is that it is very difficult to make out what such a condition is supposed to come to. Following a traditional reading of Kant's idea of transcendental limits beyond which we cannot know anything, we have perhaps a natural tendency to think of such logical limits as constituting an "outer" boundary of the sayable. But positing such bounds of sense, whether an "outer" limit of transcendental conditions beyond which we cannot think or say anything, or a lower substrate of background practices that ground what we can think or say, seems inevitably to demand the correlative notion of there being a "something" beyond or beneath the limit that we manage to grasp just long enough to say we can't think "it": this seems to require us precisely to think or talk about that which we are said to be unable to think or talk about. And this seems to me to be simply an unstable idea.[36]

The second problem with the view that coping practices function as the ground of intelligibility concerns the fact that it is simply very difficult to imagine how something conceptually normative is supposed to rest on something fundamentally alien to it. Dreyfus starts from the correct thought that an adequate phenomenology shows that humans are able to comport themselves intelligently in myriad situations without engaging in the kind of conscious deliberation favored by much of the philosophical tradition. But he wrongly concludes from this that such comportment cannot be rational at all, resting as he believes it must on motor intentionality of embodied coping skills that eschew concepts altogether.[37] Unfortunately, this exclusion of rationality and conceptuality at the level of coping skills relies on a tendentious and reified

[36] The considerations I am rehearsing here draw on what I take to be some of the most seminal writings on Wittgenstein from what have come to be known alternately as "resolute" approaches to the *Tractatus*. There is a large literature pertaining to the resolute reading of the *Tractatus*. For an overview, see Silver Bronzo, "The Resolute Reading and Its Critics: An Introduction to the Literature," *Wittgenstein-Studien*, 3 (2012): 45–80.

[37] Dreyfus uses the expression 'motor intentionality' for this kind of intelligibility, a term he takes over from Merleau-Ponty. See 'Return of the Myth of the Mental', 352, 363, and "Response to McDowell," 375–376.

understanding of both. As the following passage illustrates, what is at stake for McDowell need be nothing more than truly minimal and only roughly specifiable conditions required for the use of the demonstrative 'this'. Describing someone who catches a Frisbee on the spur of the moment, McDowell claims that while this person need not deliberate, nevertheless

> she does realize a concept of, say, catching *this*. (Think of a case in which, as one walks across a park, a Frisbee flies towards one, and one catches it on the spur of the moment.) When a dog catches a Frisbee, he is not realizing any practical concept; in the relevant sense, he has none. The point of saying that the rational agent, unlike the dog, is realizing a concept in doing what she does is that her doing, under a specification that captures the content of the practical concept that she is realizing, comes within the scope of her practical rationality—even if only in that, if asked why she caught the Frisbee, she would answer "No particular reason; I just felt like it."[38]

This passage nicely brings into view an unsatisfactory polarity in Dreyfus' understanding. On one pole, Dreyfus seems to imagine that *any* conceptuality at all necessarily draws in its train the entirety of thought and language. And because this intellectualist view of conceptuality is shown to be implausible by phenomenological inspection of the contexts in question, Dreyfus rejects conceptuality here *tout court*. At the opposite pole, the position that Dreyfus occupies, we find a close neighbor of the Augustinian picture criticized earlier. That picture commits us to openly psychologistic, brute word–object relations in which there is no room for normativity at all. Given Dreyfus' insistence that coping practices undergird and so in effect are beyond (or beneath) conceptuality, it is perhaps irrelevant whether the example concerns the seemingly simple case of "this" or any other more complex concepts. The problem with this quasi-Augustinian picture remains the same: how the "higher" level of concepts and the "lower" level of coping practices are supposed to be meaningfully integrated is utterly mysterious. Without this integration, the picture forces us to accept a bifurcation of brute alien nature on the one side and

[38] McDowell, "Response to Dreyfus," 369. Cf. G.E.M. Anscombe, *Intention*, 30–34, 72–73.

thought "spinning in a frictionless void" on the other.[39] In other words, with regard to intelligence we exhibit in nonreflective coping, we are faced with a false choice between two unsatisfactory poles: between supposing all conceptual relations must be in play or that none at all can be.

In rejecting this polarity, McDowell tries to articulate a viewpoint wherein even the apparently simplest uses of language such as a demonstrative are regarded as woven into our form of life without taking on the sort of intellectualist baggage that Dreyfus rightly suspects. On this view, it makes a qualitative difference if the world of solicitations can be, however 'roughly', taken up explicitly into the conceptual life of an acculturated, rational being. At the same time, in this 'transformative' view of rationality, our second nature not only makes us *sui generis*, there is, one might say, a 'seamless garment' in which uses of expressions such as 'this' and what might be thought of as more paradigmatic exercises of conceptuality are woven together.[40] I believe this captures McDowell's meaning when he recommends embracing 'the Hegelian image in which the conceptual is unbounded on the outside.'[41] While nonreflective coping with solicitations may retain a sort of *factual* primacy for humans when compared with more paradigmatic cases of rationality, for McDowell it loses any philosophical or *logical* priority as the supposed foundation of rationality.[42]

At one point in his APA address Dreyfus quotes Heidegger describing the character of our relation to solicitations:

> [W]hat is first of all "given" …is the "for writing," the "for going in and out," … "for sitting." That is, writing, going-in-and-out, sitting, and the like are what we are *a priori* involved with. What we know when we "know our way around."

This phrasing leads Dreyfus to worry that,

---

[39] See *MW*, 11.

[40] I have the term 'transformative' from Matthew Boyle.

[41] *Mind and World*, 83. See also *MW*, pp. 39–40, 105.

[42] See Dreyfus, "Response to McDowell," 376–377.

> Heidegger might seem to be a conceptualist himself since he continues: "My being in the world *is* nothing other than this already-operating-with-understanding."

Dreyfus immediately reassures his reader that Heidegger is not in fact a conceptualist by pointing out that Heidegger is invoking something like a distinction between knowing-how and knowing-that. Yet in a footnote shortly after this, we find this telling revelation:

> This disclosing function of perception we share with animals and infants. Heidegger, however, connects such understanding with our understanding of our identity. In that connection we should note that mere coping with affordances gives animals what Heidegger calls an environment, but not a world. To open a *world* in Heidegger's sense requires that the affordances that matter to us and draw us in depend not merely on our needs and previous experience, as with animals, but on what matters to us given our identities, and we are capable of changing our identities and so our world. This is an important difference between human beings and animals, but since we are focusing on the role of perception in giving us a background on the basis of which we can perceive objects and justify our beliefs about them, we needn't go into it here.[43]

This is a striking admission. Factorizing "identity" (i.e., meaning) and perception so cleanly as Dreyfus does here asks us to imagine conceptuality as layered on top of what is an essentially nonconceptual biological substratum. This is a "tack-on" theory of culture. But the issue that Dreyfus maintains can be put off until later discussion is in fact precisely what has been contested from the get-go and so to frame the dispute in this way begs the very question between himself and McDowell. For as we've seen, for the latter, it makes all the difference in the world that we are talking about human action and perception, and not merely certain capabilities we share with animals.[44] Now there is nothing wrong with the suggestion that practices, use, customs, etc. so to speak *carry* normativity. That is presumably why Wittgenstein implores us to look at these

[43] "Overcoming the Myth of the Mental," 65, fn 54.

[44] See "What Myth," 347.

in the first place. But the picture of these features of social life as *undergirding* the very ideas of rationality and of correctness simply doesn't work; it inherits the problems of, or perhaps simply is logically equivalent to, the Augustinian picture. Dreyfus' version of a transcendental practice reading of Wittgenstein is unsuccessful as an attempt to capture his anthropological naturalism.[45]

Culture and Evolution

In a paper written around the same time as *Mind and World*, McDowell remarked,

> When [Wittgenstein] says, "What has to be accepted, the given, is—so one could say—forms of life", his point is not to adumbrate a philosophical response, on such lines, to supposedly good questions about the possibility of meaning and understanding or intentionality generally, but to remind us of something we can take in the proper way only after we are equipped to see that such questions are based on a mistake. His point is to remind us that the natural phenomenon that is normal human life is itself already shaped by meaning and understanding.[46]

In the introduction to *Mind and World*, on the other hand, he wrote,

> Human beings acquire a second nature in part by being initiated into conceptual capacities…whose interrelations belong in the *sui generis* logical space of reasons….[47]

Now one might be forgiven for sensing a tension between the last sentence of the first passage and the entirety of the second. For it is not obvious how to square the idea that Wittgenstein aims to release us from the craving for a philosophical explanation of meaning by reminding us that our life is *already* shaped by meaning, with the second idea, that we

[45] The bald naturalist, behaviorist reading on my view doesn't even get into the game.
[46] McDowell, "Meaning and Intentionality," 50–51. Cf. Wittgenstein, *PI*, p. 226.
[47] *MW*, XX.

acquire the second nature that opens up our lives to the very phenomenon that, according to the first passage, always already shapes our life. This appears to threaten to reinstate the same tack-on picture of culture we want to avoid.

At least two questions here arise: When do we acquire our second nature and how do we acquire our second nature? Answering the first question is easy in one way. Each of us receives a second nature by growing up in an historical culture or, as the case may be, cultures. As McDowell asserts,

> Now it is not even clearly intelligible to suppose a creature might be born at home in the space of reasons. Human beings are not: they are born mere animals, and they are transformed into thinkers and intentional agents in the course of coming to maturity. This transformation risks looking mysterious. But we can take it in our stride if, in our conception of the *Bildung* that is a central element in the normal maturation of human beings, we give pride of place to the learning of language.[48]

As to the "how" question, McDowell is rightly intent on resisting any full-blown, or bald, naturalization of *Bildung*:

> Given the notion of second nature, we can say that the way our lives are shaped by reason is natural, even while we deny that the structure of the space of reasons can be integrated into the layout of the realm of law. This is the partial re-enchantment of nature that I spoke of.[49]

What McDowell says elsewhere on this topic is somewhat ambiguous. For now, note that talk of "acquiring a second nature" threatens to raise the problem of a two-tiered tack-on picture all over again. I'll take up the question of the transmissibility of culture in this final section.

In tones strikingly foreshadowing McDowell, Clifford Geertz wrote already in 1966,

[48] MW, 125 (See also MW 87–88).
[49] MW, 87–88.

> that culture, rather than being added on, so to speak, to a finished or virtually finished animal, was ingredient, and centrally ingredient, in the production of that animal itself.[50]

Geertz suggests even more trenchantly that culture has been a selective force in evolution:

> In fact, as the *Homo Sapiens* brain is about three times as large as that of Australopithecines, the greater part of human cortical expansion has followed, not preceded, the "beginning" of culture, a rather inexplicable circumstance if the capacity for culture is considered to have been the unitary outcome of a quantitatively slight but qualitatively metastatic change of the freezing-of-water sort.[51]

Geertz maintains that the human brain, already in utero, is waiting to be enculturated, or rather, in some sense, already has been subject to enculturation, or at least pre-enculturation:

> [M]an's nervous system does not merely enable him to acquire culture, it positively demands that he do so if it is going to function at all.[52]
>
> A cultureless human being would probably turn out to be not an intrinsically talented though unfulfilled ape, but a wholly mindless and consequently unworkable monstrosity. Like the cabbage it so much resembles, the Homo sapiens brain, having arisen within the framework of human culture, would not be viable outside of it.[53]

Geertz here seems to me to be getting at precisely the issue of avoiding tack-on approaches to culture. But this is not a scientific naturalist,

---

[50] "The Impact of the Concept of Culture on the Concept of Man," in Geertz, *The Interpretation of Cultures* (New York: Basic Books, 1973), 47. In his last book, Geertz claimed Wittgenstein as one of his great philosophical inspirations. See Clifford Geertz, *Available Light: Anthropological Reflections on Philosophical Topics* (Princeton: Princeton University Press, 2000), xi.

[51] Ibid., 64–65.

[52] Ibid., 68.

[53] Ibid. If enough humans were unable to attain to a certain degree of intellectual and imaginative capacities, there simply would be no concept of the human; however stinging Geertz's formulation may feel here, certain people are human in a degenerate sense.

second philosophical explanation of second nature or *Bildung*. For one thing, no mechanism is being proposed. Nor is there any talk of "bridge laws" here. Geertz is just drawing attention to undeniable, qualitatively different, "before and after" general facts of nature.

But merely avoiding tack-on theories, though admirable, seems to beg an even deeper question. If, as Geertz suggests, the concept of human being is of an organism already unavoidably ready for the transmission of culture, what about there being something to be transmissible in the first place? Even if *Lebensform*, *Bildung*, Second Nature, Culture, and of course Language, are what always already shape human life, and so must be taken for granted, we know there was a time when this was simply not so. When did what has to be accepted as the given in human life, start being that which has to be accepted as the given in human life? It isn't obvious to me that this question necessarily "contains a mistake."[54] On the one hand, McDowell is willing to admit such questions as intelligible:

> How has it come about that there are animals that possess the spontaneity of understanding? That is a perfectly good question. There was a time when there were no rational animals. Suppose we had a credible account of how forces that are intelligibly operative in nature might have led to the evolution of animals with conceptual powers.[55]

And yet, on the other hand, he warns us against reading too much into such speculation:

> But this request for an evolutionary story need not look very pressing…And in this reflection we can regard the culture a human being is initiated into as a going concern; there is no particular reason why we should need to uncover or speculate about its history, let alone the origins of culture as such. Human infants are mere animals, distinctive only in their potential, and nothing occult happens to a human being in ordinary upbringing…. Mere ignorance about how human culture might have come on the scene in the first place is hardly a plausible starting point for an argument that initiation into it must actualize an extra-natural potential in human beings.[56]

---

[54] Cf. PI §189.

[55] *MW*, 123.

[56] Ibid., 123–24.

I agree with McDowell that we ought not succumb to the temptation to give an evolutionary account of second nature, if that account is understood as the same old bald naturalist strategies. He adds,

> It would be one thing to give an evolutionary account of the fact that normal human maturation includes the acquisition of a second nature, which involves responsiveness to meaning; it would be quite another thing to give a constitutive account of what responsiveness to meaning is.[57]

But unless McDowell has in mind a very unorthodox sense of "evolutionary account", I am very uncertain whether we can actually just take things in stride in the way he encourages us. And this applies to language no less than to *Bildung*, culture, or Second Nature. I think this only serves to conceal just how truly odd the sort of non-theological reenchantment of nature McDowell advocates is to our modern sensibility. This oddness is no genuine counterargument to the case for reenchantment; there is nothing sacred about modern naturalist metaphysics.

What McDowell says about second nature fitting into an evolutionary story without becoming too tightly bound to the realm of law is admittedly not very clear. Yet in a telling footnote he comes close to confessing a certain perplexity on this issue:

> It is true, however, that the good questions we can raise in the evolutionary context come as close as good questions can to the philosophical questions I want to exorcize.[58]

If it makes no sense to account for second nature in terms of the realm of law, which presumably encompasses not only physics and chemistry, but also fields like molecular genetics and neuropsychology, it seems to me that we are simply going to have to accept the emergence of language and culture as what has to be accepted as the "given," so to speak. This will be felt by many, maybe most, to simply offend modern scientific

[57] Ibid., 124.
[58] MW, 124, fn 12.

sensibilities. Perhaps the only way to remove the sting of the loss of such scientistic faith is to point out, patiently and case by case, how utterly hopeless is the metaphysical realist fantasy that somewhere out there, whether we can make sense of it not, there's just got to be some "story" whereby the normative can be accounted for in neutral terms, if not by us, then by, what, Martians, God?

I have said little about the concept of culture and Wittgenstein's relation to it. In some of his more memorable remarks concerning culture, Wittgenstein strikes Spenglerian themes of decline.

> This book is written for those who are in sympathy with the spirit in which it is written. This is, I believe, different from that of the prevailing European and American civilisation. The spirit of this civilisation the expression of which is the industry, architecture, music, of present day fascism and socialism, is a spirit that is alien and uncongenial to the author.[59]

It is perfectly legitimate to understand "culture" this way, as referring to those more thematic aspects of human life such as music, religion, technology, and art. But culture also marks a different phenomenon, in which people's embodied, often non-thematic sense of social and even cosmic reality is anchored and defined.[60] Geertz' famous Weberian description of culture as the "webs of significance" which the animal man has both spun and in which he finds himself suspended, is open enough to capture both senses.[61] When these self-spun webs are distributed uniformly enough among a group, ethnography makes empirical sense. Where such roughly large-scale, regular distribution is lacking or spread too thinly for fieldwork, the conceptual sense of culture, as in second nature or *Bildung*, will still be in play to the degree we are dealing with individuals or small groups describable as manifesting agency at all. Irrespective, then, of the feasibility of carrying out ethnographic studies,

---

[59] Ludwig Wittgenstein, *Culture and Value*, 8. This passage is taken from a sketch for a forward to what was to be published as *Philosophical Remarks*. In the sketch, Wittgenstein invokes the Spenglerian distinction between culture and civilization, which I do not explore here.

[60] Minus the transcendental baggage, Bourdieu's *habitus* can also be thought of in this manner.

[61] "Thick Description" in *The Interpretation of Cultures*, 5.

I take "culture" to mark a logical feature of human life. If there is a "proof" that culture is "necessary" for the understanding of human beings, it is the incoherence we fall into when we falsely imagine we can talk of human action without reference to it.[62]

But Wittgenstein wrote that to imagine a language is to imagine a form of life (PI, 19), a *Lebensform*. He did not use the word "culture" here, and as I just remarked, where he does use "culture," he tends to do so in a more sweeping register. Now the German word "*Lebensform*" has undeniably biological and cultural "components," even though I am suspicious of the sense of easy factorizability that this formulation may imply. One can of course note that *Lebensform* has these two dimensions or aspects and leave things at that. But with this two-toned take on *Lebensform*, it will often be tempting to regard the biological as the more fundamental stratum, while the top, more decorative layer gets tagged the "cultural."[63] But if *Lebensform* is fundamentally biological in a sense to be *contrasted* with the cultural, then it makes little sense for Wittgenstein to write in the manner he does about *agreement* (attunement) in form of life, nor, especially as he does, harmony in judgments, tying as he does both to accord in the use of a human language.

Let's return to *PI* §241–242:

> "So you are saying that human agreement decides what is true and what is false?"—It is what human beings say that is true and false; and they agree in the language they use. That is not agreement in opinions but in form of life.
>
> If language is to be a means of communication there must be agreement not only in definitions but also (queer as this may sound) in judgments.[64]

As they simply do not speak (a human language), neither do bees, dogs, nor even whales, agree in the relevant way.[65] One might insist that

[62] This in no way commits me to the transcendental uses to which some have tried to put various culture-related concepts, especially practice, which I have argued against.

[63] I have in mind especially Stanley Cavell's essay "Declining Decline" in *This New Yet Unapproachable America: Lectures after Emerson after Wittgenstein* (Albuquerque: Living Batch Press, 1989).

[64] PI §241–242.

[65] PI §25.

animals render judgments. Maybe. But even if not always done with full-blown propositional articulation (whatever that is supposed to be), the human way of judging typically means we can give human voice to the concepts manifest in them. Even when spontaneously catching a Frisbee, normally we can even provide a reason for doing *this* when asked: "No particular reason; I just felt like it." This too falls within the space of reasons. Try that with a chimp.

We simply have no model for what a "purely" biological form of human life looks like, if that is taken to mean we can imagine a human form of life independently of the varieties of webs of significance in which self-interpreting animals sit suspended.[66] Human life itself is expressed or manifested both *across* and *through* the *many* languages, the many cultures we find on earth. Just as there is no animal that is only a mammal, or only an otter, but only individual mammals, such as individual otters, so far as we know, there are no speakers of human, only Japanese speakers, Arabic speakers, German speakers, Tamil speakers, etc. Geertz wrote presciently,

> [T]he notion that the essence of what it means to be human is most clearly revealed in those features of human culture that are universal rather than in those that are distinctive to this people or that is a prejudice we are not necessarily obliged to share.[67]

No one speaks human. Few aspire to speak Esperanto (thankfully). No one *speaks* the "common behavior of mankind," although we often find that behavior useful to communicate with people with whom we struggle to speak.[68] It would seem that explanations of meaning and understanding end where they begin, in the languages and cultures comprising the human *Lebensform*.

---

[66] PI §19.

[67] Geertz, "The Impact of the Concept of Culture on the Concept of Man," *Interpretation of Cultures*, 43.

[68] PI §206.

References

Bourdieu, Pierre. 1977. *Outline of a Theory of Practice*. Cambridge: Cambridge University Press.

Dreyfus, Hubert L. 1991. *Being-In-The-World: A Commentary on Heidegger's Being and Time, Division I*. Cambridge: MIT Press.

———. 2005. Overcoming the Myth of the Mental: How Philosophers can Profit from the Phenomenology of Everyday Expertise. *Proceedings and Addresses of the American Philosophical Association* 79: 47–65.

———. 2007a. The Return of the Myth of the Mental. *Inquiry* 50: 352–365.

———. 2007b. Response to McDowell. *Inquiry* 50: 371–377.

Geertz, Clifford. 1973. *The Interpretation of Cultures*. New York: Basic Books.

McDowell, John. 1981. Non-Cognitivism and Rule-Following. In *Wittgenstein: To Follow a Rule*, ed. S.H. Holtzman and C.M. Leich, 141–162. London: Routledge and Kegan Paul.

———. 1984. Wittgenstein on Following a Rule. In *Essays on Wittgenstein's Later Philosophy*, ed. Crispin Wright, 325–363. Dordrecht: Reidel.

———. 1992. Meaning and Intentionality in Wittgenstein's Later Philosophy. In *The Wittgenstein Legacy*, ed. Peter A. French, Theodore E. Uehling Jr., Howard K. Wettstein, 40–52. Notre Dame: University of Notre Dame Press.

———. 1994. *Mind and World*. Cambridge: Harvard University Press.

———. 2007a. What Myth? *Inquiry* 50: 338–351.

———. 2007b. Response to Dreyfus. *Inquiry* 50: 366–370.

McManus, Denis. 2008. Rules, Regression and the Background: Dreyfus, Heidegger and McDowell. *European Journal of Philosophy* 16: 432–458.

Schatzki, Theodore K. 1997. Practices and Actions: A Wittgensteinian Critique of Bourdieu and Giddens. *Philosophy of the Social Sciences* 27: 283–308.

Wittgenstein, Ludwig. 1953/1997. *Philosophical Investigations*. Ed. G.E.M Anscombe and Rush Rhees. Oxford: Blackwell.

———. 1972. *On Certainty*. Ed. G.E.M. Anscombe and G.H. von Wright, Trans. Denis Paul and G.E.M. Anscombe. New York: Harper.

———. 1976. *Lectures on the Foundations of Mathematics: Cambridge, 1939*. Ed. Cora Diamond. Ithaca: Cornell University Press.

Author Index[1]

[1] Note: Page numbers followed by 'n' refer to notes.

K. M. Cahill (ed.), *Wittgenstein on Practice*,
https://doi.org/10.1007/978-3-031-68656-6

Subject Index[1]

[1] Note: Page numbers followed by 'n' refer to notes.

K. M. Cahill (ed.), *Wittgenstein on Practice*,
https://doi.org/10.1007/978-3-031-68656-6

SPRINGER NATURE

GPSR Compliance

The European Union's (EU) General Product Safety Regulation (GPSR) is a set of rules that requires consumer products to be safe and our obligations to ensure this.

If you have any concerns about our products, you can contact us on ProductSafety@springernature.com

In case Publisher is established outside the EU, the EU authorized representative is:

Springer Nature Customer Service Center GmbH
Europaplatz 3
69115 Heidelberg, Germany

The manufacturer's authorised representative in the EU is Springer Nature Customer Service Centre GmbH, Europaplatz 3, 69115 Heidelberg, Germany. If you have any concerns regarding our products, please contact ProductSafety@springernature.com

Printed and bound by CPI Group (UK) Ltd, Croydon, CR0 4YY
01/10/2026
02225912-0013